INQUIRIES INTO CHEMISTRY
Second Edition

Michael R. Abraham
The University of Oklahoma

Michael J. Pavelich
Colorado School of Mines

WAVELAND
PRESS, INC.

Prospect Heights, Illinois

D1468589

For information about this book, write or call:

    Waveland Press, Inc.
    P.O. Box 400
    Prospect Heights, Illinois 60070
    (708) 634-0081

ISBN 0-88133-562-2

Printed in United States of America

7 6 5

CONTENTS

Section Two

Section One

GUIDED INQUIRY EXPERIMENTS

A. INTRODUCTION TO THE STUDENT

This section of the laboratory manual contains experiments called Guided Inquiry experiments. They cover many of the main areas of chemistry you will encounter in the lecture portion of the course (i.e., mass relationship in chemical reactions, heats of reactions, equilibria). As with any laboratory course, the experiments are designed to help you learn some of the laboratory techniques and procedures that scientists use to investigate nature. However, the experiments are designed to also serve two other purposes.

> To introduce you to some of the basic concepts in chemistry.

> To give you experience with some of the processes (collecting data, interpreting data, forming hypotheses and generating explanations) that a scientist uses when doing research.

Unlike some more traditional laboratory approaches, you are not expected to know anything about the subject being considered before you do the lab experiments. The purpose of the lab experiments is to introduce you to the subject area, and to give you a concrete experience with the concepts before they are discussed in a more abstract way in the lecture. Consequently many of you will be exposed to certain concepts and ideas for the first time during a laboratory activity. Use the laboratory as a learning situation where you can get your first understanding of the concepts.

Science is much more than a bunch of facts and theories. The heart of every science is research; that is, the investigation of nature. The laboratories are designed to give you experience with research. As such you will be using scientific processes such as collecting data, interpreting data, forming hypotheses, testing hypotheses and explaining results. As with any research, there are a number of explanations that will fit a set of data. You are encouraged to think independently about the data you collect. Whether you deduce the ''chemically correct'' explanation for your data will not be an issue. It is only important that your explanations or hypotheses follow logically from the data and that your reasoning is clearly stated.

In these Guided Inquiry experiments you will be given specific instructions as to what experiments to conduct and you are to answer questions about the data you collect. You should do the work in the order indicated; that is, you should answer all questions asked about one part of the experiment before you do the experimental work in the next part. Feel free to answer the questions in any way that is satisfactory to you. The ''right answer'' is any one that follows logically from the data and that you are comfortable with. You are encouraged to discuss your answers with your partner, other students in the lab or with the lab instructor. Such discussions are useful in that you may pick up new ideas and think more deeply about your own ideas. Your final lab report, however, should be written on your own.

After doing a particular Guided Inquiry experiment, your instructor may wish you to do the related Open Inquiry experiments found in Section Two.

This manual contains many appendices which you might find helpful. In particular, you should study the sections on Safety in the Laboratory and on Common Equipment and Procedures before doing your first experiment.

Michael R. Abraham
Michael J. Pavelich

PHYSICAL RELATIONSHIPS—Exp. B-1

Name_____ Lab Section_____

Lab Partner _____

I. Measure the Mass, volume (using a volumetric cylinder), length and diameter of six aluminum slugs and six brass slugs. Record the data below.

Sample #	Metal	Mass (g)	Volume (cm³)	Length (cm)	Diameter(cm)

II. A. Identify a relationship between two of the properties you measured for brass by graphing. Try to come up with an algebraic relationship that expresses the pattern you found.

B. On the same sheet of graph paper, graph the same variables for the Aluminum. What are similarities and differences in the two graphs? Offer an explanation for this.

III. Using a different combination of variables repeat the data analysis outlined in parts II.A and II.B.

UNCERTAINTY—Exp. B-2

Name_____ Lab Section_____

Lab Partner _____

I. Two weighing instruments

    A. Carefully weigh the four objects indicated by your laboratory instructor using a spring scale (weigh to nearest 0.1g) and then using a triple beam balance (weigh to nearest 0.01g). (If you have any questions about using these instruments ask your laboratory instructor.) Record your findings in the following table.

Object	Spring Scale	Triple Beam Balance
_____	_____	_____
_____	_____	_____
_____	_____	_____
_____	_____	_____

    B. Compare the similarities and differences, advantages and disadvantages of each weighing instrument.

II. Reproducing measurements (Variability in Measurement)

A. Without comparing notes beforehand, ask five other students to weigh the four objects you weighed in part I. Have them use both the spring scale and triple beam balance. When this task is finished share data and record them in the tables below.

Spring Scale

Object	Your Data from part I.A	Student 1 Data	Student 2 Data	Student 3 Data	Student 4 Data	Student 5 Data
_____	_____	_____	_____	_____	_____	_____
_____	_____	_____	_____	_____	_____	_____
_____	_____	_____	_____	_____	_____	_____
_____	_____	_____	_____	_____	_____	_____

Triple Beam Balance

Object	Your Data from part I.A	Student 1 Data	Student 2 Data	Student 3 Data	Student 4 Data	Student 5 Data
_____	_____	_____	_____	_____	_____	_____
_____	_____	_____	_____	_____	_____	_____
_____	_____	_____	_____	_____	_____	_____
_____	_____	_____	_____	_____	_____	_____

B. Calculate the average mass of each object as measured by the spring scale. Express how much each object varies from the average mass as a ± (plus or minus) number. (The value of ± is found by calculating how much each of the six measurements deviates from the average and then calculating the average of the deviations.) Show your calculations for one of the objects.

Object	Mass		
_____	_____	±	_____
_____	_____	±	_____
_____	_____	±	_____
_____	_____	±	_____

C. How would you express the precision (reproducibility) of the spring balance? What would be your reaction to someone reporting the mass of an object determined on a spring balance as 125.638g? Explain.

D. Repeat parts II.B and C for the triple beam balance.

Object	Mass		
_____	_____	±	_____
_____	_____	±	_____
_____	_____	±	_____
_____	_____	±	_____

E. Compare how the measurements made on the spring balance and on the triple beam balance differed. List as many explanations as you can for observed differences.

III. Combining Measurements.

A. Weigh each of the four objects designated by your instructor on the triple beam balance, and then weigh all four objects together. Without comparing notes ask five other students to weigh each of the four objects and combination also. When this task is finished share data and record them in the table below.

Object	Your Data	Student 1 Data	Student 2 Data	Student 3 Data	Student 4 Data	Student 5 Data
1 _____	_____	_____	_____	_____	_____	_____
2 _____	_____	_____	_____	_____	_____	_____
3 _____	_____	_____	_____	_____	_____	_____
4 _____	_____	_____	_____	_____	_____	_____
Combination _____	_____	_____	_____	_____	_____	_____

B. Calculate the maximum and minimum possible total mass for the four objects by using the individual weighings. Show your calculations below.

C. Compare the range of values (highest value − lowest value) obtained from part B with the range of values obtained for the combination weighing. Express numbers in "±" form and discuss differences and similarities. What possible ramifications does this comparison have on measurement and on calculations with measurements?

GRAPHING RELATIONSHIPS—Exp. B-3

Name_____ Lab Section_____

Lab Partner _____

I. °F vs. °C

    A. Arrange a ring stand with a ring, wire gauze, and a 250 mL beaker half filled with water. Measure a temperature of the water with a Celsius and a Fahrenheit thermometer (to ± 0.2°). Be careful to measure the temperatures at the same time and don't let the thermometers touch the glass walls of the beaker when measuring the temperature. Measure the temperature five more times after having varied the temperature by heating or cooling the water. Record the data obtained in the table.

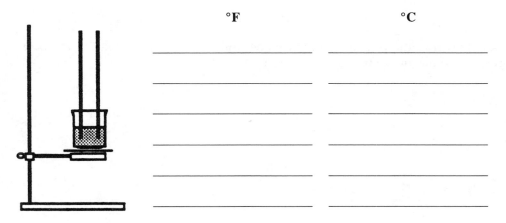

°F	°C
_____	_____
_____	_____
_____	_____
_____	_____
_____	_____
_____	_____

    B. What patterns are shown in these data? It might be helpful to graph the data. Try to come up with an algebraic equation that expresses the pattern you found.

II. Mass vs. Diameter

A. Obtain 6 rubber stoppers in different sizes from 00-13. Measure the mass of each stopper using a balance. If you are not familiar with the balance, ask your instructor to help you. Using a metric measuring stick, measure the height, top diameter, and bottom diameter of each stopper. Record your data in the table.

Stopper #	Mass (g)	Height (cm)	Top Diameter (cm)	Bottom Diameter (cm)
_____	_____	_____	_____	_____
_____	_____	_____	_____	_____
_____	_____	_____	_____	_____
_____	_____	_____	_____	_____
_____	_____	_____	_____	_____
_____	_____	_____	_____	_____

B. What patterns are shown in these data? It might be helpful to graph the data. Try to come up with an algebraic equation that expresses the pattern you found.

III. Mass vs. Volume

   A. Obtain 6 aluminum cylinders and measure each of their masses with a balance. Record the masses in the data table. Determine the volume (to $\pm$ 0.2 mL) of each aluminum cylinder by liquid displacement and record your data in the table. The liquid displacement method of determining volume is done by partially filling a 50 mL graduated cylinder with water, measuring the water, carefully submerging the aluminum in the water (tilt the cylinder and slide the aluminum down the side), and measuring the water again. The difference in the water volumes is equal to the volume of the aluminum cylinder.

Cylinder #	Mass (g)	Volume (mL)
_____	_____	_____
_____	_____	_____
_____	_____	_____
_____	_____	_____
_____	_____	_____
_____	_____	_____

   B. What patterns are shown in these data? It might be helpful to graph the data. Try to come up with an algebraic equation that expresses the pattern you found.

HYDRATES—Exp. C-1

Name_____ Lab Section_____

Lab Partner _____

I. Qualitative properties of Hydrates

    A. Place a small amount (about 2cm^3) of hydrated cupric sulfate, a small amount of hydrated ferrous sulfate and a small amount of hydrated cobalt (II) chloride into separate 150 mm Pyrex test tubes. Heat each test tube with a bunsen burner holding the test tubes at a slant (but not pointed at the person next to you!).

        1. Describe the appearance of each hydrate before heating.

        2. Describe the effects of heating on each hydrate.

    B. Add a few drops of water to each cooled test tube and record your observations.

    C. What conclusions can be drawn from these data? (e.g., What are the similarities and differences in the behavior of the three hydrates? What reaction(s) do the compounds undergo when heated? How do the data support your ideas?)

II. Quantitative Properties of Hydrated Cupric Sulfate

A. Weigh a clean, dry crucible to a precision of at least ± 0.01 g. (The proper way to prepare the crucible is to wash it, rinse with distilled water and heat for about two minutes). Add between one to four grams of hydrated cupric sulfate and record the exact weight on your data sheet. Place crucible, lid, and sample on a wire triangle on a ring stand. With the crucible lid slightly cocked, heat slowly at first, then gradually increase the heat. Do not allow the crucible to become red hot at any time. Occasionally lift the lid with tongs to observe when the dehydration is complete. When the reaction appears to be completed, continue heating for five more minutes. Let the crucible, lid, and contents cool completely, then reweigh and record the results.

B. If time permits, repeat part II.A with a different sized sample. Collect four sets of data from other students. Make sure you have a range of different weights from 1 to 4 grams.

Hydrated Cupric Sulfate Data

Before Heating	(a)	(b)	(c)	(d)	(e)	(f)
1. mass of crucible + lid	_____g	_____				
2. mass of crucible + lid + sample	_____	_____				
3. mass of sample	_____	_____	_____	_____	_____	_____

After Heating	(a)	(b)	(c)	(d)	(e)	(f)
4. mass of crucible + lid + sample	_____	_____				
5. mass of anhydrous salt	_____	_____	_____	_____	_____	_____
6. mass of water lost	_____	_____	_____	_____	_____	_____

C. What patterns are shown in these data? It might be helpful to graph the data. Try to come up with an algebraic equation that expresses the pattern you found. Explain why you chose the particular algebraic equation.

D. What generalization about mass (weight) relationships in chemicals can be drawn from the data? Specifically, what would you predict would be the results if hydrated cobalt chloride was treated as in part II.A? Explain your reasoning.

III. Empirical Formula and Equations

A. The general formula of hydrated cupric sulfate can be written in $CuSO_4(H_2O)_x$, where x represents the relative number of water molecules that are combined with each ionic unit of salt. Determine the value of x (i.e., determine the empirical formula of the hydrated salt) from the data in part II.B.

B. Using the value of x determined above, write a balanced chemical equation for the reaction that occurred when the hydrate was heated. Include in the equation the physical state of the individual chemical substances. Briefly explain the reasoning used in determining the empirical formula and chemical equation from the data.

PRECIPITATES—Exp. C-2

Name_____ Lab Section_____

Lab Partner _____

Pre-Lab Assignment: Study Appendix D.5 on filtration techniques.

I. Qualitative Properties of Cobalt Nitrate-Sodium Phosphate Interactions.

    A. Dissolve a small amount (about 1 cm³) of solid cobalt nitrate, $Co(NO_3)_2$, in about 20 mL of distilled water. Dissolve a similar amount of sodium phosphate, Na_3PO_4, in a second 20 mL of water. Describe the appearance of each solution.

    B. Pour half of each solution into a third beaker and mix thoroughly. (Save the mixture and two solutions for later use.)

        1. Describe the appearance of the mixture.

        2. Assuming the reaction involves the coming together of dissolved ions, what are the possible identities of the solid formed in the mixture? Think up and carry out experiments which would distinguish among all of the possibilities. Describe the results of these experiments.

3. Write a chemical equation which represents the reaction and is consistent with the data obtained so far. Briefly explain your reasoning.

C. Separate the mixture by filtration (see Appendix D.5). Note the characteristics of the liquid, called the supernatant, and predict all of the materials which might be dissolved in the liquid.

D. Divide the supernatant in half and test each half with the remaining $Co(NO_3)_2$ and Na_3PO_4 solutions. Describe the results.

E. What conclusions can be drawn from these data concerning the chemicals present in the supernatant? (e.g., How might changing the original amounts of $Co(NO_3)_2$ and Na_3PO_4 affect the composition of the supernatant?)

II. Quantitative properties of the $Co(NO_3)_2$—Na_3PO_4 interaction.

A. Obtain 75 mL of stock $Co(NO_3)_2$ solution in a clean, labeled beaker. Obtain 60 mL of stock Na_3PO_4 solution in a second labeled beaker. Record the concentrations on page 24.

Clean and distilled water rinse three 100 mL beakers and label them 1-3. Into each beaker measure 20.0 mL of stock $Co(NO_3)_2$ solution. Measure various amounts of stock Na_3PO_4 into each beaker. The suggested amounts are:

4.0, 12.0, and 20.0 mL **or** 8.0, 16.0, and 24.0 mL.

Mix each solution thoroughly and allow it to stand for at least 10 minutes. Calculate the masses of salts added to each beaker and record these amounts in the table on page 24.

B. Label and weigh three sheets of filter paper, recording their masses in the table. Mix and filter each of the three reaction mixtures (see Appendix D.5). Catch the first 15 - 20 mL of supernatant from each filtration in separate, clean and labeled beakers. Set these aside for later analysis in part III.A.

C. Continue the filtrations into any convenient containers. Then wash each precipitate with two 5 mL portions of water. Use these water washes to transfer any precipitate remaining in the original beakers to the filter paper.

(These washings will proceed slowly. During this time you can run the tests in part III.A, page 25, on the supernatants you saved.)

D. After the washings, put the filter papers with their filtrates in labeled beakers and then in a 110°C drying oven. Dry the filtrates for 20 or more hours. (Your instructor may have special instructions concerning the drying procedure.) Let cool and weigh. Record the data in the table. Collect nine other data sets from other students. (Your instructor will probably make special arrangements for the class to collect such data.)

Data Table
(All masses in grams)

Concentration of stock Na_3PO_4 solution _____ g/mL

Concentration of stock $Co(NO_3)_2$ solution _____ g/mL

Solution	mL Stock Na_3PO_4	mL Stock $Co(NO_3)_2$	g Na_3PO_4	g $Co(NO_3)_2$	Mass of Paper	Mass of ppt. & Paper	Mass of ppt.
1	_____	_____	_____	_____	_____	_____	_____
2	_____	_____	_____	_____	_____	_____	_____
3	_____	_____	_____	_____	_____	_____	_____
Other Students' Data	_____	_____	_____	_____	_____	_____	_____
	_____	_____	_____	_____	_____	_____	_____
	_____	_____	_____	_____	_____	_____	_____
	_____	_____	_____	_____	_____	_____	_____
	_____	_____	_____	_____	_____	_____	_____
	_____	_____	_____	_____	_____	_____	_____
	_____	_____	_____	_____	_____	_____	_____
	_____	_____	_____	_____	_____	_____	_____
	_____	_____	_____	_____	_____	_____	_____
	_____	_____	_____	_____	_____	_____	_____

III. A. Describe the appearance of each of your supernatants. Divide each of the supernatants into two parts. Test one part with a dropperful of $Co(NO_3)_2$ and the other with a dropper full of Na_3PO_4 solution. Summarize the results of these tests below.

B. Explain the result of this testing of the supernatants. That is, explain what these results show must have happened in each reaction mixture. (For example: why can you form more precipitate from a supernatant? Why don't all three supernatants give the same test results? Could you have predicted these test results from your balanced equation and masses of salts added?)

IV. A. (Optional) What patterns are shown in the mass data on page 24? Graph the mass of sodium phosphate against the mass of the precipitate and explain what is occurring in each section of the curve. How does this fit what you would predict the mass of the precipitates to be?

B. (Optional) In part I.B. you proposed an equation for the reaction you observed. Do you feel the supernatant and mass data (on pages 24 and 25) are evidence for or against your equation being correct? Briefly explain your reasoning.

ZINC AND HYDROCHLORIC ACID—Exp. C-3

Name_____ Lab Section_____

Lab Partner _____

Pre-lab assignment: Review your textbook discussion of molar (mole/liter) concentration units and study Appendix D.1 on the use of pipets and burets.

CAUTION: In this experiment you will be dealing with fairly concentrated solutions of acids and bases. These substances are very corrosive, especially to the human body. Handle these substances with extreme care. If any spills occur, wash and wipe up the spill immediately. If you get any on your person, wash the affected area generously with water. See your instructor if a burning sensation is felt.

I. Qualitative Investigation.

A. Obtain three chunks of zinc (mossy or shot form) and about 100 mL of 5M hydrochloric acid solution.

Put about 50 mL of distilled water in a 100 mL beaker and add one chunk of Zn. Record your observations.

Put about 50 mL of 5M HCl in a 100 mL beaker and add one chunk of Zn. Record your observations.

B. You will now run an experiment to trap the gas given off in the reaction so that it can be tested and identified. Fill a test tube with water and invert it into a 100 mL beaker of water such that no air is trapped in the test tube.

Adjust the water level in the beaker so that it just covers the mouth of the test tube. Be careful that no air enters the test tube. Add about 50 mL of 5M HCl and mix the solution. Add a fresh chunk of Zn and position the test tube such that it captures most of the gas being formed.

When the test tube is filled with gas, pull it from the solution keeping it in a vertical position with open end down. Let the flame of a match lick over the opening of the tube. Record your observations.

C. Speculate on the reaction that is occurring. Write a chemical equation to represent the reaction. Briefly explain why you chose this equation. (See Table C-2 in Appendix C.)

II. A Quantitative Investigation

A. Clean and distilled water rinse five 250 mL Erlenmeyer flasks. Label them 1 through 5. Obtain about 75 mL of a 5M HCl solution (The beaker used should be cleaned and rinsed with two 5 mL portions of the 5M HCl solution). Record the exact molarity of the HCl solution in the data table on page 30.

Rinse a 10 mL pipet with two 2-3 mL portions of the HCl solution (be sure to use a pipetting bulb to draw the solution into the pipet, see Appendix D.1). Pipet exactly 10.00 mL of the HCl solution into each of the five Erlenmeyer flasks. Calculate the number of moles of HCl delivered to each flask and record the values in the data table.

Using small beakers or weighing papers, weigh out two samples of *granular* zinc (10 to 20 mesh). Each sample should have a mass between 0.1 and 0.5 grams. Record the exact masses in the table under trials 4 and 5. Add these Zn samples slowly to Erlenmeyer flasks #4 and #5, respectively. If the reaction becomes too violent (bubbling reaching half way up the flask), stop the addition until the reaction subsides. When all the Zn has been added cover the two flasks with watch glasses and set them aside until the reactions are complete (30 to 40 minutes). About every 10 minutes swirl the flasks to mix the reactants. In the meantime proceed with the practice titrations in parts II.B and C.

28

B. The purpose of the remainder of this experiment is to determine the number of moles of HCl that remain in flasks 4 and 5 after the Zn reaction is complete. You will do this by adding enough NaOH solution to just consume the remaining HCl. This analysis technique is called titration. The analysis reaction is:

$$HCl\ (aq)\ +\ NaOH\ (aq)\ \longrightarrow\ H_2O\ +\ NaCl\ (aq)$$

You will use the dye methyl orange to indicate when the analysis is complete. The dye changes color from pink to orange at the point where just enough NaOH has been added to consume the HCl. This point is called the endpoint of the titration.

At this point # moles HCl = # moles NaOH added

You will first titrate the HCl solutions in flasks #1 to 3 to obtain experience with recognizing the methyl orange color change and to obtain baseline data.

C. Obtain about 100 mL of stock NaOH solution (the beaker should be rinsed with two 5 mL portions of the solution to prevent contamination or dilution of the solution). Record the exact molarity of the NaOH solution in the data table. Rinse and fill a buret with the NaOH solution according to the directions in Appendix D.1. Record the initial buret reading in the table.

Add 2 to 4 drops of methyl orange indicator solution to flask #1 and swirl to mix. Slowly drain the NaOH solution from the buret into the flask, swirling the flask after each addition to mix the reactants. Continue this process until one drop of NaOH solution changes the indicator to a permanent yellow to yellow orange color. Record the final buret reading in the data table.

Calculate the total volume of NaOH solution used, the number of moles of NaOH used and the number of moles of HCl consumed by the NaOH. Record each of these in the table. Titrate the HCl solutions in flasks #2 and 3, calculate the quantities asked for and compare the results from the three solutions for reproducability. If the results are not within 1% of each other, talk to the instructor about ways to improve your experimental technique, or reread Appendix D.1.

D. When the reactions in flasks 4 and 5 are complete, wash down the watch glass and sides of the flask with small amounts of distilled water to return splattered material to the solution. Add 2 to 4 drops of methyl orange solution to each of the flasks and carefully titrate them with NaOH solution. (Localized white precipitate, $Zn(OH)_2$, will form where the NaOH first hits the solution. Swirl the flask until the precipitate disappears before adding the next amount of NaOH. If more than a few seconds of swirling is needed to dissolve the precipitate, you are close to the end point and the NaOH should be added a drop or two at a time.) Record the titration data in the table. Calculate the number of moles of NaOH added and number of moles of HCl titrated for each case. Collect three other sets of data on Zn reaction solutions from other students.

DATA TABLE

Molarity of stock HCl _____ M

Molarity of stock NaOH _____ M

Trial

	1	2	3	4	5	other students' data		
# moles HCl added:	_____	_____	_____	_____	_____	_____	_____	_____
# grams Zn added:	0.00	0.00	0.00	_____	_____	_____	_____	_____
Initial buret reading (mL):	_____	_____	_____	_____	_____	_____	_____	_____
Final buret reading (mL):	_____	_____	_____	_____	_____	_____	_____	_____
Volume of NaOH used (mL):	_____	_____	_____	_____	_____	_____	_____	_____
# moles NaOH added:	_____	_____	_____	_____	_____	_____	_____	_____
# moles HCl titrated:	_____	_____	_____	_____	_____	_____	_____	_____

E. Show your calculations for one of the trials below.

III. A. What patterns are shown in the data on pg. 30? It might be helpful to graph the data. Find an algebraic equation that expresses this pattern. Explain why you chose this algebraic equation.

B. Offer a physical explanation for the pattern in the data. That is, discuss on a qualitative level why the amount of HCl after reaction with Zn varied as it did.

IV. (Optional) In part I.C on page 28 you speculated on what balanced chemical equation best represented the reaction you observed. Show mathematically how the data on page 30 can be used to confirm or deny the chemical equation. Explain your reasoning.

SPECTRAL ANALYSIS FOR Cu²⁺(aq)—Exp. C-4

Name_____ Lab Section_____

Lab Partner _____

Pre-lab assignment: Review textbook discussion of Molar (moles/liter) concentration units. Study Appendix D.1 on pipets and burets and Appendix D.2 on the theory and use of spectrophotometers.

I. A Qualitative Look at a Reaction.

A. Dissolve about 1 cm³ of hydrated copper (II) sulfate, $CuSO_4(H_2O)_5$, in 25 mL of distilled water. Dissolve about 1 cm³ of KOH pellets in another 25 mL of water. (**Caution**: KOH is a strong base and very corrosive to metal and flesh. Handle it with care). Describe the appearance of each solution.

B. Pour the solutions together and describe the results.

C. Assuming the reaction involves the combining of dissolved ions, what are the possible identities of the precipitate formed in the mixture? Devise and conduct experiments that will distinguish between the possibilities. Describe the results of these experiments.

D. Write a chemical equation which represents this reaction and is consistent with the data obtained so far.

II. A Quantitative Investigation of the Reaction.

You will be making quantitative measurements on the reaction of KOH(aq) with $CuSO_4$(aq) by using a spectrophotometer. This instrument will measure the color intensity and thus the concentrations of Cu^{2+}(aq), the ion responsible for the blue color of the solutions. Since the experiments depend on concentrations, it is important for the precision of the results that all volume and mass measurements are made with care. Review Appendix D.1 on volume measurement techniques. Also, you should be careful to not unwittingly dilute or contaminate your solutions.

A. Obtain about 120 mL of stock Cu^{2+}(aq) solution and record its concentration on page 35. This solution was prepared by dissolving a weighed amount of $CuSO_4(H_2O)_5$ to a known volume with H_2O.

B. Label two clean and dried 100 mL beakers. Into each weigh 0.5 to 1.8 g of KOH. Record the exact masses in the table on page 35. Add 25.00 mL of distilled water to each beaker. When the KOH has completely dissolved add 25.00 mL of the stock Cu^{2+}(aq) solution to each beaker with gentle mixing. Record the exact volumes used in the table. Let these solutions stand while you do part II.C. However they should be stirred periodically. (To prevent contamination or loss of chemicals it is best to leave the stirring rods in their respective beakers.)

C. Prepare four different solutions having known concentrations of Cu^{2+}(aq). This is most conveniently done by diluting parts of your stock Cu^{2+}(aq) solution. The following dilutions of Cu^{2+}(aq) stock to distilled H_2O are suggested: 10.00 mL: 10.00 mL, 10.00 mL: 15.00 mL, 10.00 mL: 20.00 mL, 10.00 mL: 30.00 mL. Record the actual make-up of your diluted solutions in the Table. Calculate the mole/liter concentration of each solution and record it in the table.

D. Prepare filter paper cones for two filter funnels (see Appendix D.5.) **Do not** wet the papers with water as this would cause a small dilution of the reaction solutions. Filter the two reaction solutions into separate, labeled, dry beakers. You are to obtain a pure sample of the supernatant (liquid part of the mixture). It may be necessary to pass the solution through the filter two or three times to obtain a sample clear of precipitate. It is only important that about 15 mL of the pure supernatant is collected. The filter paper and precipitate can be discarded.

E. You are now ready to measure the color intensity (Absorbance) of blue Cu^{2+} in each of your solutions. Follow the instructions in Appendix D.2 and/or those provided by the instructor. It is suggested that the wavelength of light used for the measurements be 620 nm.

Stock Cu^{2+}(aq) solution concentration _____ M.

Reaction Solutions	Trial #1	Trial #2
mass of beaker (g)	_____	_____
mass of beaker + KOH	_____	_____
mass of KOH	_____	_____
mL H_2O added	_____	_____
mL stock Cu^{2+} added	_____	_____
total mL solution	_____	_____
Absorbance at 620 nm	_____	_____

Standard (Diluted) Cu^{2+}(aq) solutions

	A	B	C	D
mL stock Cu^{2+}	_____	_____	_____	_____
mL H2O added	_____	_____	_____	_____
Conc. of Cu^{2+}(M)	_____	_____	_____	_____
Absorbance at 620 nm	_____	_____	_____	_____

III. A. Analysis of the Absorbance Data for Standard Cu^{2+}(aq) Solutions.

Find a pattern between the concentrations and the absorbances of your standard (diluted) Cu^{2+}(aq) solutions. It may be helpful to graph the data. Find an algebraic equation that expresses this pattern, evaluate all constants in your equation and determine their units. Show and explain your work below.

B. The Analysis of the Reaction Solutions.

Considering the algebraic equation and/or graph established in part III.A, determine the molar concentration of Cu^{2+}(aq) in each of your reaction solutions from their absorbances. Enter these values in the table on page 37. Make the other calculations called for in the table. Show your work for reaction Solution #1 in the space below. Complete the table by recording other students' data.

	Reaction Solution #1	Reaction Solution #2	Other Students' Data				
Concentration of Cu^{2+} after reaction	_____	_____	_____	_____	_____	_____
moles of Cu^{2+} after reaction	_____	_____	_____	_____	_____	_____
moles of Cu^{2+} before reaction	_____	_____	_____	_____	_____	_____
moles of Cu^{2+} consumed	_____	_____	_____	_____	_____	_____
mass of KOH added	_____	_____	_____	_____	_____	_____
moles of KOH added	_____	_____	_____	_____	_____	_____

IV. A. What patterns are shown in the data above? Develop an algebraic equation that expresses the pattern you found. Explain why you chose the particular algebraic equation.

B. In part I.D, page 33, you speculated on the nature of the reaction between KOH(aq) and CuSO₄(aq). Use the data on page 37 to prove or disprove this speculation. Explain your reasoning below.

B. In part I.D, page 33, you speculated on the nature of the reaction between $KOH(aq)$ and $CuSO_4(aq)$. Use the data on page 37 to prove or disprove this speculation. Explain your reasoning below.

CuSO₄ AND KI—Exp. C-5

Name_____ Lab Section_____

Lab Partner _____

Pre-lab assignment: Review your textbook discussion of molar concentration units and study Appendix D.1 on the use of pipets and burets.

I. Qualitative Investigation.

   A. Dissolve about 1 cm³ of solid hydrated copper sulfate, $CuSO_4(H_2O)_5$, in 30 mL of distilled water. Dissolve about 1/2 cm³ of potassium iodide, KI, in a separate 30 mL of water. Describe the appearance of each solution.

   B. Pour each solution into a third beaker and mix thoroughly. Describe the appearance of the mixture.

   C. Filter the mixture saving both residue and supernatant. Set aside at least 15 mL of the supernatant. Wash the residue two or three times with 5 mL portions of water, these washings need not be kept.

      While the washing is going on, proceed with parts D and E below.

   D. Determine which of the ions you mixed together are the reactants in this reaction. For example: what are the four ions with which you started? which sets of ions might have reacted? Think up and carry out experiments that will distinguish among the possibilities. Describe the results of these experiments.

39

E. State which ions you think are the reactants, and briefly explain your logic.

F. Describe the results of the filtration and washings done in part I.C. Do these results indicate that there are one or two product species? Briefly explain your reasoning.

G. Add several drops of the supernatant to 5 mL of cyclohexane in a large test tube. Mix vigorously and describe the results. What chemical do the results show to have been present in the filtrate (see table C-2 in Appendix C)?

H. Consult the Inorganic Compounds section of a CRC Handbook of Chemistry and Physics to determine the possible identities of the residue in the filter paper. List logical candidates and their published characteristics.

J. What are the products of the reaction? Briefly describe your logic.

K. Write a balanced chemical equation for the reaction.

II. A Quantitative Investigation

A. Precisely weigh between 0.2 and 0.5 grams of $CuSO_4(H_2O)_5$ into a clean 250 mL Erlenmeyer flask. Record the exact mass in the data table on the next page. Add about 30 mL of distilled water and dissolve the copper salt.

Roughly weigh 2.5 to 3.0 grams of KI into a small beaker, add 20 mL of water to dissolve it, and then add 10 mL of 0.18 M acetic acid. Pour this mixture into the Erlynmeyer flask containing $CuSO_4$. Swirl and let stand for a few minutes to ensure that the reaction is complete.

B. Obtain about 50 mL of $Na_2S_2O_3$ solution in a clean labeled beaker. Take precautions to not dilute or contaminate your sample of this solution. Record its exact concentration in the data table. Prepare a buret and fill it with the solution as described in Appendix D.1.

You are to titrate the $Na_2S_2O_3$ into your reaction mixture. The thiosulfate ion reacts with the I_2 in the Erlenmeyer flask according to the equation:

$$I_2 + 2S_2O_3^{2-} \longrightarrow 2I^- + S_4O_6^{2-}$$

However, since iodine will react with our excess iodide as:

$$I_2 + I^- \longrightarrow I_3^-$$

the actual titration reaction is:

$$I_3^- + 2S_2O_3^{2-} \longrightarrow 3I^- + S_4O_6^{2-}$$

The titration end-point occurs when you have just dispelled the color of the I_2. At the end-point your solution will be a milky white. We will add starch near the end to help locate the end-point.

Titrate your reaction mixture until it turns from a dirty brown to a milky brown. Add 2 mL of starch solution, swirl to mix, and continue to titrate slowly until the blue of the starch-iodine complex is dispelled. Record the volume of thiosulfate used in the data table.

C. If time allows, titrate another reaction mixture using a different weight of Cu salt. In any case, gather enough data from other students to complete the data table.

Concentration of $Na_2S_2O_3$ = _____ M

Titration Data

g of $CuSO_4(H_2O)_5$ _____ _____ _____ _____ _____

mol of Cu^{2+} _____ _____ _____ _____ _____

initial buret
reading _____ _____ _____ _____ _____

final buret
reading _____ _____ _____ _____ _____

mL of $Na_2S_2O_3$ _____ _____ _____ _____ _____

mol of $S_2O_3^{2-}$ _____ _____ _____ _____ _____

mol of I_2 (I_3^-) _____ _____ _____ _____ _____

III. In Part I.K you wrote a balanced chemical equation for the reaction between ions in $CuSO_4$ and KI. Show mathematically how your titration data confirm or deny this equation. Explain your reasoning. If the equation does not fit the data, find a new equation that does. Show explicitly that this new equation does fit both the qualitative and titration data.

DISSOLUTION REACTIONS—Exp. D-1

Name_____ Lab Section_____

Lab Partner _____

I. Qualitative Properties of Some Dissolving Reactions

A. Place about 25 mL of distilled water into a 50 mL beaker and suspend a 0.1°C thermometer into the liquid using a ring stand and a thermometer clamp. (Take care that the thermometer is not touching the bottom of the beaker as any movement of the beaker could snap the thermometer.) Place a moderate amount (approximately 1 to 3 cm³) of Anhydrous Magnesium Sulfate, $MgSO_4$, into the water, mix vigorously for about 15 seconds and record your observations.

Repeat this procedure using fresh distilled water with each of the following compounds and record your observations.

B. Sodium Nitrate, $NaNO_3$

C. Sodium Chloride, $NaCl$

D. Hydrated Calcium Chloride, $CaCl_2(H_2O)_2$

E. Ammonium Nitrate, NH_4NO_3

F. What conclusions can be drawn from these data? (e.g., What are the similarities and differences in the behavior of the several compounds? Does the data indicate any generalizations concerning all chemical reactions?)

II. Quantitative properties of the reaction $\quad MgSO_4(s) \xrightarrow{H_2O} Mg^{2+}(aq) + SO_4^{2-}(aq)$

A. Using a beaker accurately weigh a 3 to 8 gram sample of $MgSO_4$ on the analytical balance. Record the exact mass in the table on page 49.

B. Suspend a 0.1°C thermometer into a polystyrene cup so that it is about 1/2 inch from the bottom. Using a volumetric cylinder, add 100.0 mL of distilled water and stir while recording time versus temperature data in the table. (Data should probably be taken every 30 sec.) After about 4 minutes add the $MgSO_4$ with vigorous mixing while continuing to record data. Mixing and the recording of data should continue for 5 to 10 minutes after the $MgSO_4$ has dissolved.

C. Determine ΔT, the temperature change for the reaction, from your data and explain how you did it. (i.e., Graph your data by plotting temperature versus time, draw the best curved line through the points, and explain what is occurring in each section of the curve. What part of the curve represents ΔT for the dissolution reaction?)

D. Calculate the heat, Q, of the reaction from the equation $Q = CM \, \Delta T$
assume $C = 4.18$ joules/gram °C and M is the mass of the water (take the water density as 1.00 grams/cm³).

E. Repeat the experiment using a different sized sample of $MgSO_4$. Show your analysis of the data below. Collect four more sets of data from other students. Record all the data on the Data Summary Table, page 50.

Data Sheet

Trial #1	Trial #2

a. mass of beaker _____   _____

b. mass of beaker _____   _____
   +MgSO$_4$

c. mass of MgSO$_4$ _____   _____

Time	Temp	Time	Temp
_____	_____	_____	_____
_____	_____	_____	_____
_____	_____	_____	_____
_____	_____	_____	_____
_____	_____	_____	_____
_____	_____	_____	_____
_____	_____	_____	_____
_____	_____	_____	_____
_____	_____	_____	_____
_____	_____	_____	_____
_____	_____	_____	_____
_____	_____	_____	_____
_____	_____	_____	_____
_____	_____	_____	_____
_____	_____	_____	_____
_____	_____	_____	_____
_____	_____	_____	_____
_____	_____	_____	_____
_____	_____	_____	_____
_____	_____	_____	_____
_____	_____	_____	_____
_____	_____	_____	_____

Summary of Data

	#1	#2	#3	#4	#5	#6
a. Mass of $MgSO_4$	_____	_____	_____	_____	_____	_____
b. Mass of water	_____	_____	_____	_____	_____	_____
c. ΔT	_____	_____	_____	_____	_____	_____
d. Heat, Q	_____	_____	_____	_____	_____	_____

F. What pattern is shown by the data above? It might be helpful to graph the data. Give an algebraic equation that expresses this pattern. What conclusions or generalizations can be obtained from the data? Explain all your reasoning.

POTASSIUM HYDROXIDE AND HYDROCHLORIC ACID
Exp. D-2

Name_____ Lab Section_____

Lab Partner _____

CAUTION!!!

You will be working with the strong base potassium hydroxide (KOH) and the strong acid hydrochloric acid (HCl). These chemicals are extremely corrosive to metals and human flesh. If small amounts of the chemicals are spilled, immediately wipe up the spill, wash the area with water and wipe it dry. If the chemicals get on your person, wash the affected area immediately under a stream of water. If large spills occur, have the instructor handle it.

Do not place the chemicals directly on the balance pans, weigh the chemicals in beakers.

I. The Reaction: $KOH(s) \xrightarrow{\text{H}_2\text{O}} KOH \text{ (aq)}$

A. Weigh between 2.5 and 5 grams of KOH pellets into a small, clean and dry beaker. Record the exact mass in the table on page 53.

Suspend a $0.1\,°C$ thermometer into a polystyrene cup so that it is about 1/2 inch from the bottom. Using a graduated cylinder, add 100.0 mL of distilled water and stir while recording time versus temperature data in the following table. (Readings should be taken every 30 sec.) After about 3 minutes add the sample of solid KOH while continuing to stir vigorously and record data. Data should be collected at 30 sec. intervals for the next five minutes.

When data taking is completed, pour the KOH solution into a clean, dry labeled beaker. Cover this with a watch glass to prevent contamination. This solution will be used in part III.

B. Graph your time versus temperature data. From the graph determine the temperature change, ΔT, for the dissolving reaction. Explain how you did this. Record the ΔT value on page 53.

Reaction #1

mass of beaker (grams) _____

mass of beaker + KOH _____

mass of KOH _____

Time	Temp
_____	_____
_____	_____
_____	_____
_____	_____
_____	_____
_____	_____
_____	_____
_____	_____
_____	_____
_____	_____
_____	_____
_____	_____
_____	_____
_____	_____
_____	_____
_____	_____
_____	_____
_____	_____

ΔT for the reaction_____

II. The Reaction: KOH(s) + HCl(aq) ⟶ H₂O + KCl(aq)

A. Weigh out another 2.5 to 5 g sample of KOH. Record the mass on page 55. Suspend the 0.1°C thermometer as before into your clean, dry polystyrene cup.

B. Secure about 175 mL of 1.5 M HCl solution in a labeled beaker. Record the exact concentration of the solution in the table on page 55. Using a volumetric cylinder put 100.0 mL of the HCl solution into the cup. Stir and record time versus temperature data as before. After about 3 min. add the KOH and continue to stir and record data for 5 min. Test the final solution with pH paper.

C. Graph these data and determine ΔT for the reaction. Briefly explain your reasoning. Was there a limiting reagent in this reaction? Briefly explain your reasoning.

III. The Reaction: KOH(aq) + HCl(aq) ⟶ H₂O + KCl(aq)

A. Rinse your graduated cylinder with distilled water then measure out 50.0 mL of the KOH solution from part I into a small beaker. Rinse the cylinder and measure 50.0 mL of the 1.5 M HCl into your dry polystyrene cup. Calculate the weight and mole information called for in the table on page 55. Rinse and dry the end of the thermometer. Measure the temperature of the KOH solution and record it.

B. Rinse and dry the thermometer then suspend it in the HCl solution. Stir and record time versus temperature data for two minutes. Add the 50.0 mL of KOH solution while stirring. Record time versus temperature data for about 3 minutes. Test the final solution with pH paper.

C. Graph these data and determine ΔT for the reaction. Briefly explain your reasoning. (If the initial temperatures of the two solutions are different, how should this be handled in determining ΔT for the reaction?) Was there a limiting reagent in the reaction? Briefly explain your reasoning.

Concentration of HCl solution (M) _____

Reaction #2 (KOH solid)   **Reaction #3** (KOH solution)

mass of beaker _____   # grams KOH in 50 mL_____

mass of beaker + KOH_____   # moles HCl of 50 mL _____

mass of KOH _____   initial temp. of KOH solution _____

moles HCl in 100 mL_____

Time	**Temp**	**Time**	**Temp**
_____	_____	_____	_____
_____	_____	_____	_____
_____	_____	_____	_____
_____	_____	_____	_____
_____	_____	_____	_____
_____	_____	_____	_____
_____	_____	_____	_____
_____	_____	_____	_____
_____	_____	_____	_____
_____	_____	_____	_____
_____	_____	_____	_____
_____	_____	_____	_____
_____	_____	_____	_____
_____	_____	_____	_____
_____	_____	_____	_____
_____	_____		

pH test result _____   pH test result _____

ΔT for the reaction _____ ΔT for the reaction _____

IV. A. Compare the time versus temperature graphs for the three reactions. What differences exist among them? Offer a brief explanation for these differences.

B. Transfer your data to the proper table on page 57. Calculate the number of moles and heat, Q, called for. The heat is found by the equation: $Q = C M \Delta T$

As an approximation take $C = 4.18$ joules/g°C and M as the mass of water (assume there is 1.0g of H_2O for every 1.0 mL of solution). Show your calculations of Q below.

C. Complete the tables on page 57 by collecting data from other students.

I. $KOH(s) \xrightarrow{\text{H}_2\text{O}} KOH(aq)$

    a. mass of KOH(g) _____ _____ _____ _____ _____

    b. moles of KOH _____ _____ _____ _____ _____

    c. mL of solution _____ _____ _____ _____ _____

    d. $\Delta T(^\circ C)$ _____ _____ _____ _____ _____

    e. heat, Q(J) _____ _____ _____ _____ _____

    f. heat/moles KOH (J/m) _____ _____ _____ _____ _____

II. $KOH(s) + HCl(aq) \longrightarrow H_2O + KCl(aq)$

    a. mass of KOH(g) _____ _____ _____ _____ _____

    b. moles of KOH _____ _____ _____ _____ _____

    c. moles of HCl _____ _____ _____ _____ _____

    d. mL of solution _____ _____ _____ _____ _____

    e. pH test _____ _____ _____ _____ _____

    f. limiting reagent _____ _____ _____ _____ _____

    g. $\Delta T(^\circ C)$ _____ _____ _____ _____ _____

    h. heat, Q(J) _____ _____ _____ _____ _____

    i. heat/moles KOH (J/m) _____ _____ _____ _____ _____

III. $KOH(aq) + HCl(aq) \longrightarrow H_2O + KCl(aq)$

    a. mass of KOH(g) _____ _____ _____ _____ _____

    b. moles of KOH _____ _____ _____ _____ _____

    c. moles of HCl _____ _____ _____ _____ _____

    d. mL of final solution _____ _____ _____ _____ _____

    e. pH test _____ _____ _____ _____ _____

    f. limiting reagent _____ _____ _____ _____ _____

    g. $\Delta T(^\circ C)$ _____ _____ _____ _____ _____

    h. heat, Q(J) _____ _____ _____ _____ _____

    i. heat/moles KOH (J/m) _____ _____ _____ _____ _____

D. How do the data you collected on the individual reactions compare with the data collected by other students? (Identify a pattern between the amount of KOH and Q(or ΔT) in the data for each reaction individually. Find algebraic expressions for these patterns.)

E. Identify a relationship among the heats of the three reactions. (For instance, is there a mathematical relationship among the chemical equations? Is there a similar relationship among their heats?) Explain your reasoning. Try to find a statement in your text that expresses the same relationship you found in the data.

HEATING AND COOLING BEHAVIOR—Exp. D-3

Name_____ Lab Section_____

Lab Partner _____

I. Cooling Lauric Acid

A. Obtain a sample of between 12-15 g of Lauric Acid in an 18 x 150 mm Pyrex test tube. Remove the stopper and place the test tube in a 400 mL beaker half-filled with tap water. Put the beaker on a ring stand and heat the water with a bunsen burner. Do not let the water boil violently.

B. Immerse a thermometer into the sample when it becomes liquid. When the sample reaches a temperature between 65-70° C, remove the test tube from the hot water and clamp it to a ring stand. Precede immediately to part C.

C. Make a water bath by filling a 400 mL beaker 3/4 full of tap water at room temperature (about 25°C).

Hold the thermometer in the center of the liquid sample in the test tube. Do not allow the thermometer to touch the bottom of the test tube. Note the temperature of the sample and the time. Immediately immerse the test tube into the water bath. Continue noting the temperature of the sample every 30 seconds while observing the state of the sample. Record your data in the table. Continue recording the temperatures until the sample is in the 30-35°C range.

D. What patterns are show in these data? It might be helpful to graph the data.

Time	Temperature	Observations
____	____	_____
____	____	_____
____	____	_____
____	____	_____
____	____	_____
____	____	_____
____	____	_____
____	____	_____
____	____	_____
____	____	_____
____	____	_____
____	____	_____
____	____	_____
____	____	_____
____	____	_____
____	____	_____
____	____	_____
____	____	_____
____	____	_____
____	____	_____
____	____	_____
____	____	_____

II. Heating Lauric Acid

A. Allow the sample and the test tube from part I to cool to room temperature (about 25°C). The thermometer should be embedded in the sample.

Make a hot water bath by putting 300 mL of water into a 400 mL beaker and heating it until the temperature of the water is approximately 70°C. Turn off the burner, but leave it in position. Note the temperature of the solid sample, and the time; and then immediately immerse the test tube into the water bath. Clamp the test tube into position. Do not allow the test tube to touch the bottom of the water bath. Continue noting the temperature of the sample every 30 seconds, observing the state of the sample until the temperature of the sample is in the 60°C range. Record your data in the table. As soon as the solid sample breaks free of the sides of the test tube, gently mix the sample with the thermometer. If the temperature of the water bath falls below 60°C, heat it until the temperature is approximately 65°C.

B. What patterns are shown in these data? It might be helpful to graph the data.

C. Compare and contrast the heating lauric acid data with the cooling lauric acid data obtained in part I. What do the data demonstrate about lauric acid?

Time	Temperature	Observations
____	____	_____
____	____	_____
____	____	_____
____	____	_____
____	____	_____
____	____	_____
____	____	_____
____	____	_____
____	____	_____
____	____	_____
____	____	_____
____	____	_____
____	____	_____
____	____	_____
____	____	_____
____	____	_____
____	____	_____
____	____	_____
____	____	_____
____	____	_____
____	____	_____

III. Heating Water.

A. Fill a 400 mL beaker half full of water at room temperature. Clamp a thermometer so that its tip is immersed about half of the depth of the water. Note the temperature of the water and the time and immediately begin heating the water with a burner. Continue noting the temperature of the water every 30 seconds while observing the state of the sample. Record your data in the table. Continue recording the temperature for about 10 minutes (or at least three minutes after boiling begins).

B. What patterns are shown in these data? It might be helpful to graph the data.

C. The data patterns (zones) found in parts I, II, and III can be generalized to all substances. Sketch a theoretical time/temperature graph for water being heated from ice at $-10°C$ to steam at $120°C$ and discuss what each zone represents.

Time	Temperature	Observations
_____	_____	_____
_____	_____	_____
_____	_____	_____
_____	_____	_____
_____	_____	_____
_____	_____	_____
_____	_____	_____
_____	_____	_____
_____	_____	_____
_____	_____	_____
_____	_____	_____
_____	_____	_____
_____	_____	_____
_____	_____	_____
_____	_____	_____
_____	_____	_____
_____	_____	_____
_____	_____	_____
_____	_____	_____
_____	_____	_____
_____	_____	_____
_____	_____	_____

IV. Sketch a time/temperature graph for the heating of lauric acid from room temperature (25°C) to 150°C. Useful information concerning lauric acid can be found in reference books (i.e., *Handbook of Chemistry and Physics*).

V. Speculate on why the graph you sketched in part IV is shaped as it is? (For example, what might the added heat be doing to the molecules in each section of the graph?)

PRESSURE, VOLUME, TEMPERATURE RELATIONSHIPS
Exp. E-1

Name_____ Lab Section_____

Lab Partner _____

I. Pressure and Volume

A. Obtain a pressure measuring device (manometer) as shown in figure A.

Figure A

B. Obtain a 30 mL syringe and fill it with air to the largest volume mark on the syringe. Connect the syringe to the manometer at the short hose and test for gas leaks. If you cannot eliminate all leaks, see your lab instructor. Observe the levels of the mercury column in both arms of the manometer. What is the pressure of the trapped air in the syringe? Explain.

C. Depress the plunger of the syringe all the way and describe the system. Is the pressure of the trapped air greater or less than atmospheric pressure? Explain.

D. Release the plunger of the syringe. The syringe plunger should be resting somewhere between the 25 and 30 mL markings. You might want to rotate the plunger in the syringe to eliminate any sticking. Compress the trapped air by pushing on the plunger until the mercury levels are no longer equal. Record the value for the volume in the syringe and the height of the left and right mercury columns in the following table. Take volume and height readings for five more compressions down to a syringe volume of 0 mL. Obtain atmospheric pressure. Record these values in the following table.

Atmospheric pressure = _____mm Hg at_____ (time),_____ (date)

Hg height of Syringe Arm	Hg height of open Arm	Syringe Volume
_____	_____	_____
_____	_____	_____
_____	_____	_____
_____	_____	_____
_____	_____	_____
_____	_____	_____

E. Calculate the total pressure of the trapped air for each reading and record it in the following table. Show how you calculate this pressure for your first reading in the space below.

F. Calculate the total volume of the trapped air for each reading and record it in the following table. Show how you calculated this volume for your first reading. (Hint: treat the volume in the syringe arm of the manometer as a cylinder, $V = \pi r^2 L$.)

Data Table

Pressure	Volume
_____	_____
_____	_____
_____	_____
_____	_____
_____	_____
_____	_____

G. What patterns are shown in these data? It might be helpful to graph the data. Try to come up with an algebraic equation that expresses the pattern you found. Explain why you chose the particular algebraic equation.

II. Pressure and Temperature

A. Assemble a 250 mL Erlenmeyer flask as shown in figure B. Connect this to the long hose of the manometer and test for gas leaks.

Figure B—Gas Container

B. Using a meter ring and wire gauze, support a 1000 mL beaker half way up a ring stand. Remove the rubber stopper from the flask. Using a clamp suspend the flask in the 1000 mL beaker. The flask should be positioned so that only its neck is above the beaker rim and does not touch the sides or the bottom of the beaker.

Fill the beaker with tap water to about 1 1/2 cm from the rim. Be careful to **not** get any water into the flask.

Using a gas burner, heat the water to a temperature of 75-80°C. The water should be constantly stirred during the heating process. A thermometer can be used as the stirring rod; however, care must be taken to not break the fragile thermometer.

When the temperature has reached 75-80°C, remove the heat and immediately connect the flask to the long hose of the manometer. While continuing to gently stir the water allow the temperature to drop about 10°C.

C. Check to make sure the rubber stopper is firmly placed in the flask. Record the temperature and the height of Hg in both arms of the manometer in the following table.

Allow the water to cool approximately another 5°C (continue stirring the water). Record the temperature and the exact height of Hg in both arms of the manometer in the following table.

Using the procedure outlined above continue to record readings at approximately 5°C intervals until a temperature of about 15°C is reached.

Notes:

1. If the stopper at any time comes out of the bottle the experiment must be restarted.
2. The cooling can be hastened by adding small amounts of ice to the water bath. However, to insure the gas sample has reached the temperature of the water, stir for at least 3 minutes after the ice has disappeared before taking readings.
3. Excess water can be removed from the water bath. However, the water level should be at most 3 cm from the rim.

Data Table

Height of Hg in Manometer		Temperature (°C)	Pressure of Trapped Gas
Open Arm	Flask Arm		
_____	_____	_____	_____
_____	_____	_____	_____
_____	_____	_____	_____
_____	_____	_____	_____
_____	_____	_____	_____
_____	_____	_____	_____
_____	_____	_____	_____
_____	_____	_____	_____
_____	_____	_____	_____
_____	_____	_____	_____
_____	_____	_____	_____
_____	_____	_____	_____
_____	_____	_____	_____
_____	_____	_____	_____

D. Calculate the actual pressure of the trapped gas as you did in part I.E and record the information on the preceding table. Show the calculation for one of the readings below.

E. What pattern is shown by the data? It might be helpful to graph the data. Give an algebraic equation that expresses this pattern. Explain your reasoning.

III. Extention of Gas Relationships

A. (Optional) Develop a single algebraic expression which relates the three parameters of a gas which you have studied (pressure, temperature, and volume). Show how this expression is consistent with the expressions you have developed in parts I and II. Use this expression to predict what would happen to the pressure of a gas sample if its volume and temperature were decreased simultaneously.

B. (Optional) If you could continually cool a gas sample, estimate the temperature at which its pressure is reduced to zero. Discuss the significance of this temperature.

PRESSURE OF VAPORS—Exp. E-2

Name_____ Lab Section_____

Lab Partner _____

I. The Apparatus

A. Obtain a manometer, a clean and dry 125 mL suction flask, and the other materials needed to set up the apparatus shown in Figure A.

Figure A

B. Fill the 1000 mL beaker with tap water. Obtain the temperature and atmospheric pressure and record these data in the following table. Open the stopcock and turn on the aspirator until no further change takes place. Close the stopcock and then turn off the aspirator. Test the apparatus for leaks. If no change occurs for several minutes proceed. If not, see your instructor.

C. Describe what happened. Compare the manometer reading with atmospheric pressure.

D. Calculate the pressure in the flask. Show your calculations below and record the result in the following table.

II. Hexane (**Caution**: hexane is a flammable liquid. Keep it away from open flames.)

A. Disconnect the aspirator and open the stopcock. Open the suction flask and add 15 mL of hexane. Reconnect the apparatus, close the stopcock, and turn on the aspirator. Open and close the stopcock until a maximum height of mercury is attained. Close the stopcock and turn off the aspirator. Make sure some of the liquid remains in the flask. Wait 5-10 minutes until the temperature stabilizes and the mercury column reading becomes constant. Record the temperature and the heights of both arms of the manometer in the following table. Compare the readings with the readings obtained when the flask was empty. Propose a reason for any differences noticed.

B. Cool the flask by adding ice to the 1000 mL beaker. When the ice has melted and the temperature is stabilized, record the temperature and the height data in the following table. Continue cooling and recording until you have 6 readings in all.

Data Table

Atmospheric pressure _____ at _____ (time), _____ (date)

Pressure inside empty flask_____ at_____ (temp)

Temperature	Open Arm	Flask Arm

C. Disconnect the aspirator, open the stopcock, pour out the hexane, connect the system and evacuate it dry.

D. Calculate the pressure exerted by the vapor of the hexane for each temperature measured and record these data in the following table. Show the calculations for the first data point below.

Data Table

Temperature	Vapor Pressure (Hexane)
_____	_____
_____	_____
_____	_____
_____	_____
_____	_____
_____	_____

E. What patterns are shown in these data? It might be helpful to graph the data.

III. Methanol and Acetone

A. Repeat part II with methanol and acetone. (**Caution**: methanol and acetone are flammable liquids. Keep them away from open flames.)

Methanol

Open Arm	Flask Arm	Pressure	Temperature
_____	_____	_____	_____
_____	_____	_____	_____
_____	_____	_____	_____
_____	_____	_____	_____
_____	_____	_____	_____
_____	_____	_____	_____

Acetone

Open Arm	Flask Arm	Pressure	Temperature
_____	_____	_____	_____
_____	_____	_____	_____
_____	_____	_____	_____
_____	_____	_____	_____
_____	_____	_____	_____
_____	_____	_____	_____

B. Discuss and compare the vapor pressure plots for the three liquids. Propose a theory to explain why these liquids behave as they do. How would you expect water to behave in comparison with the three liquids?

CHEMICAL PROPERTIES—Exp. F-1

Name_____ Lab Section_____

Lab Partner _____

I. General Instructions

A. **CAUTION**: Most of the chemicals you will be using can cause severe burns to flesh and clothing. Handle all solid chemicals with forceps, tongs or a spatula, **NOT WITH YOUR FINGERS**. Be particularly careful to not let **Na or K** metal come in contact with your person.

Take care that none of the solutions you create or take from the reagent shelf come into contact with your person or are spilled on the lab bench. If this does occur rinse the affected area generously with water.

B. Chemical Facts Needed

1. H_2 gas will burn in air, O_2 and Cl_2 gas will not. O_2 gas will support burning. A glowing splint thrust into O_2 will burst into flame. Cl_2 gas has a choking odor. (See Table C-2 in Appendix C.)
2. An acid is any compound that will release H^+ ions in aqueous solution. A base is any compound that will release OH^- ions in aqueous solution. An amphoteric compound is one that will act either as an acid or as a base depending on what it reacts with.
3. An acidic compound that is soluble in water will turn blue litmus pink, will not affect pink litmus and will give a colorless solution when phenolphthalein is added. A water soluble base will not affect blue litmus paper, will turn pink litmus blue and will give a red solution when phenolphthalein is added. A basic compound that is insoluble in water will often dissolve in an acidic (i.e., HCl) solution. An acidic compound that is insoluble in water will show the opposite behavior.

II. Reactions with Water

A. Put about 150 mL of water into a 250 mL beaker. Fill a Pyrex test tube with water and invert it into the beaker so that little or no air is trapped in the top of the test tube. Using forceps or tongs loosely wrap a piece of sodium metal in aluminum foil. Punch several holes in the foil to expose the lump of Na. Using forceps or tongs immerse the wrapped sodium into the beaker of water and collect the evolved gas in the test tube. When the test tube is completely filled with gas pull it out of solution being careful to keep it in a vertical position with the open end down. Touch a lighted match to the open end and record your observations. Test the solution you have made with litmus paper or phenolphthalein and record your observations.

Write a balanced chemical equation for the reaction you have observed.

B. Repeat the experiment above using a chip of calcium metal (it need not be wrapped), and record your observations.

Write a balanced chemical equation for the reaction you have observed.

C. Explain how your observations support the chemical equations you have written.

D. In this experiment you are to observe the relative vigor with which various elements react with water. Fill three beakers about half way with water, cover the beakers with watch glasses. Into one beaker place a piece of Na metal, K metal into another and Ca metal into the third. Record your observations.

Repeat this procedure with a piece of Mg metal, Al metal and a pinch of sulfur. Record your observations.

E. Is there any correlation between the reactivity of these elements and their position in the periodic table? Explain your answer. (i.e., what would you predict would occur if rubidium (Rb) metal was placed in water? If chlorine was placed in water? How do your observations support your prediction?)

III. Reactions with Acid

A. Put 75 mL of water into a 250 mL beaker, fill a Pyrex test tube with water and invert it into the beaker as you did in part II.A. Add about 25 mL of 4.0M HCl solution to the beaker and mix. Using tongs, hold 2-3 pieces of Mg in the solution collecting the gas in the test tube. Test the gas with a match as you did in part II.A. Record your observations.

Write a balanced chemical equation for the reaction you have observed.

Explain the reasoning you used in coming up with this equation.

B. Place 5-10 mL of 4.0M HCl solution into each of three test tubes. Drop a piece of Mg metal into one test tube, Al metal into another and a pinch of sulphur into the third. Record your observations.

C. Is there a correlation between reactivity with acid and the position of the element in the periodic table? Explain your answer. (i.e., what would you predict would happen if sodium or chlorine were placed in acid solution? Why do you predict this?)

IV. Acid-Base Properties

In these experiments you will be dealing with compounds having the general formula EO_xH_y where x and y will be various whole numbers and E represents various elements. You are to determine whether these compounds are acid, base, neutral, or amphoteric compounds. (See part I.B.)

A. Dissolve a pellet of NaOH in about 50 mL of water. Test the solution to determine its acid-base properties. Explain how you did this and record your results.

B. The compound MgO_2H_2 has been added to water for you. The compound is relatively insoluble in water. Shake the solution bottle and obtain about 5 mL of the slurry. Test the slurry to determine the acid-base properties of MgO_2H_2. Explain how you did this and record your results.

C. The compounds COH_4, PO_4H_3, SO_4H_2, and ClO_4H have been dissolved in water for you also. Obtain about 5 mL of each solution. Test each for acid-base properties and record your observations and conclusions.

COH_4

PO_4H_3

SO_4H_2

ClO_4H

D. (Optional) Put 7 mL of 1 M $Al(NO_3)_3$ into a Pyrex test tube and add to it 3 mL of 6 M NH_4OH. Gently heat the test tube with a bunsen burner for about 2 minutes (do not point the test tube at anyone). Allow the tube to cool, centrifuge it for about 3 minutes, then pour off the liquid. Fill the test tube halfway with distilled water, mix the solid with the water, centrifuge and again pour off the liquid. The solid you have formed is AlO_3H_3.

With a stirring rod transfer a small amount of the solid to two test tubes. Add about 2 mL of 4 M HCl a drop at a time to one test tube and about 2 mL of 4 M NaOH a drop at a time to the other. Record your observations and your conclusions concerning the acid-base properties of AlO_3H_3.

E. Write the formulas of the ions that are present when the following compounds are dissolved in water.

NaOH

SO_4H_2

ClO_4H

COH_4

F. What conclusion can be drawn about the acid-base properties of EO_xH_y compounds and the position of the element, E, in the periodic table?

V. Select one of the trends you have observed in this experiment and develop an explanation for it.

PROPERTIES OF HALOGENS—Exp. F-2

Name_____ Lab Section_____

Lab Partner _____

I. Solubility of the Halogens (Cl_2, Br_2, and I_2) and the Halides (Cl^-, Br^-, and I^-).

    A. Test the solubility of Cl^-, Br^-, and I^- in both water and cyclohexane (C_6H_{12}) by adding a small amount (enough to cover a match head) of NaCl, NaBr, and NaI to separate test tubes and adding 0.5-1.0 mL of water. Repeat the test with 0.5-1.0 mL of C_6H_{12}. Mix vigorously. Record your observations. What conclusions can you draw?

    B. Mix the Cl^- (in water) with the Cl^- (in C_6H_{12}). Mix the solutions vigorously for 15 seconds. Do the same for the Br^- solutions and I^- solutions. Record your observations and conclusions.

C. Test the solubility of a few I_2 crystals in water and in C_6H_{12} in a manner similar to part I.A. Add about 1.0 mL of C_6H_{12} to the water solution of I_2 and mix vigorously. Record your observations and conclusions. Contrast I^- with I_2.

D. Obtain 10 mL each of water solutions of I_2, Br_2, and Cl_2. Place 5 drops of each in separate test tubes and add about 1 mL of C_6H_{12}. Mix vigorously. Repeat the test with 10 drops of each in about 1.0 mL of C_6H_{12}. Record your observations.

E. Summarize your observations on the colors and solubility of the halides and halogens in the following chart.

	Cl^-	Br^-	I^-	Cl_2	Br_2	I_2
in water						
in cyclohexane						

II. Interactions between the halogens and halides

   A. Add 5 drops of Br_2 (in water) to 1 mL of Cl^- (in water). Mix thoroughly. Add 0.5 mL of C_6H_{12} and mix vigorously. Record your observations.

   Does the cyclohexane test indicate a reaction between Br_2 and Cl^-? If so, write a balanced equation describing the reaction. If there was no reaction write no reaction. Defend your conclusion using the experimental data you collected.

   B. Repeat the experiment substituting I^- (in water) for the Cl^-. Record your observations. Write and defend your equation.

C. Repeat parts II.A and B using Cl_2 (in water) with Br^- (in water) and with I^- (in water). Record your observations, write and defend your equations.

D. Repeat parts II.A and B using I_2 (in water) with Cl^- (in water) and with Br^- (in water). Record your observations, write and defend your equations.

III. Using the Data

A. Using the data collected in part II compare the reactivity of the halogens. Compare the reactivity of the halides. Is there any correlation between the reactivity of the halogens and halides and their position in the periodic table? Explain your answer. (i.e., what would you predict would occur if F_2 and At_2 were reacted with Br^-? Why do you predict this?)

B. (Optional) If you did experiment F-1 compare the results obtained in this experiment with those obtained in that experiment.

ACID AND BASE CLASSIFICATIONS—Exp. G-1

Name_____ Lab Section_____

Lab Partner _____

I. Acid/Base Characteristics

A. Obtain a 96 well microplate. With a medicine dropper or dropper bottle carefully 1/2 fill each well in column 1 (rows A-F) with 1 M NaOH solution (see diagram).

B. Do the same with columns 2-7 with 1 M HCl, 1 M H_2SO_4, 1 M HNO_3, saturated $Ca(OH)_2$, 1 M KOH, and distilled water respectively. Rinse the dropper when changing solutions.

C. Dip small pieces of red and blue litmus paper in each of the solutions in row A (see diagram) and record your observations in the table.

		NaOH	HCl	H_2SO_4	HNO_3	$Ca(OH)_2$	KOH	H_2O
		1	2	3	4	5	6	7
Litmus	A	○	○	○	○	○	○	○
BTB	B	○	○	○	○	○	○	○
PHN	C	○	○	○	○	○	○	○
Mg	D	○	○	○	○	○	○	○
$CaCO_3$	E	○	○	○	○	○	○	○
$Mg(NO_3)_2$	F	○	○	○	○	○	○	○

D. Add one microdrop of bromothymol blue (BTB) to each of the solutions in row B and one microdrop of phenolphthalein (PHN) to each of the solutions in row C. Record your observations in the table.

E. Place a small piece of magnesium metal in each of the solutions in row D. Record your observations in the table.

F. Place a small amount (an amount just covering the tip of a spatula) of solid $CaCO_3$ in each of the solutions in row E. Record your observations in the table.

G. Add one microdrop of $Mg(NO_3)_2$ solution to each of the solutions in row F. Record your observations in the table.

H. Test each of the seven solutions with a light bulb conductivity apparatus. To do this fill a 100 mL beaker half full of distilled water. Add 10 mL of the test solution. Test each with the conductivity apparatus. Record your observations in the table.

	NaOH	HCl	H_2SO_4	HNO_3	$Ca(OH)_2$	KOH	Distilled Water
Litmus							
Bromothymol blue							
Phenolphthalein							
Mg							
$CaCO_3$							
$Mg(NO_3)_2$							
Conductivity							

I. Group the seven solutions according to similar properties. What are the least number of groups needed? What substances are in each group?

J. Using their chemical formulas, identify similarities and differences among each of the groups from part I.

K. Write an equation for any one of the reactions you observed when you added the $Mg(NO_3)_2$ solution.

Obtain 20 mL of 1.0 M HCl and divide it into two 50 mL beakers. Put several pieces of Mg into one solution and cover the beaker with a watch glass. After waiting several minutes and without removing the watch glass hold a lighted match to the pouring spout of the beaker. Record your observations. Write a chemical equation which represents the reaction.

Using the same procedure outlined above put several chips of $CaCO_3$ into the second sample of 1.0 M HCl and test with a lighted match. Record your observations and write a chemical equation.

L. Assume the HCl is one of a class of compounds called acids and that NaOH is one of a class of compounds called bases. Summarize what you learned about acids and bases in this investigation so far.

II. Concentrations of Acids and Bases

A. Obtain 5 mL of each of these solutions and put them in separate clean test tubes: 6M HCl, 1 M HCl, and 0.1 M HCl. Add a single chip of $CaCO_3$ to each solution. Describe and explain the results of this experiment.

B. Put 10 mL of a 0.1 M HCl solution in a clean test tube and label it "10^{-1} M H^+," using a 10 mL graduated cylinder or a 10 mL graduated pipet transfer 1 mL of the 10^{-1} M solution and 9 mL of distilled water to a second test tube mix and label it "10^{-2} M H^+." Rinse and "shake dry" your measuring equipment. Transfer 1 mL of the 10^{-2} M solution to a third test tube, add 9 mL of distilled water mix and label this "10^{-3} M H^+." Continue this "serial dilution" until you have 5 solutions from 10^{-1} M to 10^{-5}M.

C. Obtain 10 mL of a 0.1 M NaOH solution and carry out a "serial dilution" similar to part II.B. At the end you should have 5 solutions labeled from "10^{-1}M OH^-" to "10^{-5}M OH^-."

D. Obtain a strip of wide range pH paper. Dip a clean dry glass rod into a sample of distilled water and then touch the rod to a small section of the pH paper. Compare the color of the paper with the color code provided with the paper and record the value in the table below. Using the same procedure, test the ten solutions you made in parts II.B and C. Record these data in the table below.

Distilled water pH = _____

Acid		Base	
Dilution	**pH**	**Dilution**	**pH**
10^{-1}	_____	10^{-1}	_____
10^{-2}	_____	10^{-2}	_____
10^{-3}	_____	10^{-3}	_____
10^{-4}	_____	10^{-4}	_____
10^{-5}	_____	10^{-5}	_____

E. What conclusions can be drawn from these data?

III. Comparisons of Different Acids and Bases

A. Obtain 5 mL of each of 6 M acetic acid and 1 M acetic acid ($HC_2H_3O_2$). Add a single chip of $CaCO_3$ to each solution. Compare the rate of these reactions to those of part II.A. Repeat the II.A reactions if a more direct comparison is necessary.

How can these data be accounted for? Discuss the similarities and differences between HCl and $HC_2H_3O_2$.

B. Obtain a few mL of each of the following solutions: 0.1 M HCl, 0.1 M HNO_3, 0.1 M H_2SO_4, 0.1 M $HC_2H_3O_2$, and 0.1 M H_3BO_3. Test each solution with pH paper using the procedure outlined earlier. Record your observations and account for the similarities and differences among the 5 acids.

IV. Acid/Base Interactions

   A. Using the graduations printed on a 50 mL beaker add 0.1 M NaOH solution to the 10 mL mark. Add two drops of bromothymol blue to this solution. Note the color of the solution. Titrate this solution with 0.1 M HCl solution (by adding HCl from a dropper bottle dropwise, swirling the solution during addition) until one drop of the HCl changes the color of the solution. Note the color and the approximate amount of HCl added. Write an equation for the reaction.

   B. Repeat the experiment using 0.1 M HNO_3 instead of 0.1 M HCl. Compare your findings with part IV.A.

   C. Repeat the experiment using 0.1 $HC_2H_3O_2$ instead of HCl or HNO_3. Compare your findings with parts IV.A and B.

D. What conclusions can be drawn from these data? Compare these data with the information gathered concerning pH of these acids in part III.B. If you had used 0.05 M HCl in part IV.A how would the results have been different? (feel free to try this by diluting the HCl solution)

E. (Optional) Repeat the experiment of part IV.A, B, and C using 0.1 M H_2SO_4 instead of HCl, HNO_3, or $HC_2H_3O_2$. Compare your data with that of those sections and account for any differences.

ACID AND BASE INTERACTIONS—Exp. G-2

Name_____ Lab Section_____

Lab Partner _____

Pre-lab assignment: Study Appendices D.1 and D.3.

CAUTION!!!

In this laboratory we will be investigating and using some of the properties of acids and bases. You will work in pairs to facilitate data collection. Due to the possible injurious effects of acids and bases on both metallic equipment and the human body, the following precautions should be followed at all times:

1. Never allow the solid acids and bases or their solutions to come into contact with your person or clothing. If such contact does occur wash the area profusely with cold water. Wear your safety glasses at all times.

2. If a solid or solution should spill, wipe up the spill immediately and rinse the area and wipe cloth with water.

3. Do not set chemicals directly on the analytical balance pans, use preweighed beakers or flasks to hold the chemicals.

I. Obtain 10-12 mL of 6 M NaOH solution and add it to 500 mL of distilled water and thoroughly mix the solution. Prepare a storage container by thoroughly cleaning it, rinsing it twice with distilled water and then rinsing it twice with 5 mL portions of your NaOH solution. Pour about 300 mL of your NaOH solution into the container, stopper it tightly and store in your drawer.

II. Because of a limited number of pH meters, it is suggested that some groups do part C of this section first. Since parts A, B, and C can be done in any order, this should cause no difficulties. However, you should read over part A before proceeding to part C.

A. Into a clean, distilled water-rinsed 250 mL Erlenmeyer flask, precisely weigh about 0.5 grams of $HKC_8H_4O_4$, potassium acid phthalate (commonly abbreviated KHP, mole weight = 204.2 g/mole). Record the exact mass on your data sheet. Add 50 mL of distilled water and swirl until all the acid is dissolved. Add 2 drops of phenolphthalein solution and swirl to mix.

Rinse a 50 mL buret twice with approximately 10 mL portions of NaOH solution. Fill the buret with NaOH solution making sure the stem is filled with liquid and contains no air bubbles. Record the volume reading of the meniscus. Slowly allow the NaOH solution to drain into the acid solution. The acid flask should be swirled after each addition of NaOH solution. When the acid solution turns a permanent pale pink or red color the titration is finished. Record the volume reading of the meniscus and calculate the total volume of NaOH added.

Repeat this procedure using approximately 0.7 grams of potassium acid phthalate weighed accurately. It is expected that the partners will alternate doing the complete experiment on the different sized samples.

B. Repeat the titration procedure in part A on a 0.3 and 0.6 gram sample of potassium acid phthalate using 2 drops of bromothymol blue instead of phenolphthalein.

C. Repeat the titration procedure in part A on a 0.4 gram sample of potassium acid phthalate using a pH meter instead of either phenolphthalein or bromothymol blue. (You may, however, add one of these to the solution to check the pH of the changing color.)

You should use two mL increments in the beginning of the titration (pH=3 to pH=5), dropwise increments in the middle (pH=4 to pH=9), and one mL increments in the end (pH=9 to pH=12).

IN USING THE pH METER BE SURE TO FOLLOW THE DIRECTIONS GIVEN TO YOU BY YOUR INSTRUCTOR TO THE LETTER!

III. (Optional) If time allows repeat the titration procedure in part II.A using 0.3 - 0.7 g of potassium acid phthalate and 2 drops of methyl orange instead of phenolphthalein.

DATA

	Phenolphthalein		Bromothymol Blue		Methyl Orange (optional)
Titrations with Dyes	1	2	1	2	
mass of flask	_____	_____	_____	_____	_____
mass of flask + KHP	_____	_____	_____	_____	_____
mass of KHP	_____	_____	_____	_____	_____
# of moles of KHP (HKC$_8$H$_4$O$_4$)	_____	_____	_____	_____	_____
initial NaOH volume reading	_____	_____	_____	_____	_____
final NaOH volume reading	_____	_____	_____	_____	_____
total NaOH volume used	_____	_____	_____	_____	_____

Titration with pH Meter

buret readings	pH reading
0.00 mL	_____
_____	_____
_____	_____
_____	_____
_____	_____
_____	_____
_____	_____
_____	_____
_____	_____
_____	_____
_____	_____
_____	_____
_____	_____
_____	_____
_____	_____
_____	_____
_____	_____
_____	_____
_____	_____
_____	_____

mass of flask   _____

mass of flask + KHP   _____

mass of KHP   _____

moles of KHP   _____

Volume of NaOH at the end-
point (pH = 9 for KHP)*   _____

* In general the end-point is taken as the point of steepest rise in the pH versus mL of NaOH curve on a graph.

IV. Analyzing the Data

A. Make a graph of the pH-meter titration data by plotting pH vs. mL of NaOH solution added. (Traditionally pH is put on the Y-axis and mL is on the X-axis.) Determine the end point of this titration and explain how you did this.

B. What patterns are shown in the weight vs. end point volume data among the phenolphthalein, bromothymol blue and pH titrations? (It may be useful to graph the data.)

Derive a mathematical expression for this pattern. How might the behavior you observed be explained? Be sure to point out how your explanations are supported by the observations.

C. What are the similarities and differences in the three methods of part I? What is the difference in using bromothymol blue and phenolphthalein? What are the advantages and disadvantages of using a pH meter?

V. (Optional) How well does the titration using methyl orange fit the pattern established by the other titrations? Offer an explanation of this methyl orange result.

VI. One important use of titration experiments is that they can be used to determine the exact concentration of solutions. The concentration of the NaOH solution prepared in part I cannot be accurately calculated from molarity and dilution used to make up the solution. These numbers are not that well known and NaOH contains impurities. However, the acids that were titrated are reasonably pure and accurately measured; therefore, the concentration of the NaOH solution can be calculated from the titration data.

Calculate the concentration of your NaOH solution from each set of titration data (excluding that with methyl orange), then determine an average value for [OH⁻] in that solution. The information on the following page may be helpful in making these calculations.

Titration	# moles KHP	# moles H^+ in KHP	# moles OH^- added	# liters OH^- added	[OH^-]
1	_____	_____	_____	_____	_____
2	_____	_____	_____	_____	_____
3	_____	_____	_____	_____	_____
4	_____	_____	_____	_____	_____
5	_____	_____	_____	_____	_____

Average _____

VII. Reactions involving acid base interactions generally take the form:

Acid + Base ⟶ Water + Salt

Write a balanced equation for the reaction studied here. Label all of reactants and products (i.e. identify the salt).

By definition:

(1) conc. of OH^- solution = $\dfrac{\text{\# moles of } OH^-\text{added}}{\text{\# liters of } OH^-\text{solution}}$

The numerator of the right hand term can be determined using the following reasoning:

Each molecule of the acid KHP can release one H^+ ion when dissolved in water according to the following equation:

Therefore:

(2) # moles H^+ available in the compound = # moles KHP. The net reaction that is occurring as you titrate is: $H^+ + OH^- \longrightarrow H_2O$. The OH^- added is consuming the H^+ being released by the compound. What is important is that at the end point of the titration:

(3) # moles OH^- added = # moles H^+ available in the compound.

MOLECULAR STRUCTURES—Exp. H

Pre-lab assignment: Study the sections of your textbook on Lewis Structures and Molecular Geometries (i.e., Valence-Shell Electron-Pair Repulsion theory). Alternatively or in addition, you can study the Supplements to this experiment on pages 114-127.

Study Supplement #3, page 128 on 3-D drawings.

In this laboratory period you will be asked to determine the Lewis Structures and the three dimensional geometries of molecular species starting only with molecular formulas. Thus there will be no lab work, per se. The lab time is to be used to help you perfect your skills at deducing Lewis Structures and, most importantly, to help you become acquainted with various molecular geometries. Model kits or balls and sticks will be available so that you can construct 3-D models of the molecules.

You will be assigned to work with one of the lists of molecular species on the next page. You are to do the following for each species, writing your answers on the work sheets provided.

1. Determine the number of valence electrons in the molecule. (See your text and/or Supplement #1, page 115.)

2. Determine the Lewis Structure of the molecule. (See your text and/or Supplement #1, page 116.)

3. Determine the geometry around each centralized atom in the molecule from the Lewis Structure. (See your text and/or Supplement #2, page 120.)

4. Construct an exact geometric model of the molecule using the materials provided.

5. Draw an exact 3-D representation of the molecule from the model. (See Supplement #3, page 128.)

Examples of how the work sheets are to be filled in are given on page 112.

MOLECULAR SPECIES LIST

List #1

1. $SiCl_4$ (no Cl-Cl bonds)
2. PCl_3 (no Cl-Cl bonds)
3. NO_3^- (no O-O bonds)
4. SF_6 (no F-F bonds)
5. PCl_5 (no Cl-Cl bonds)
6. IF_3 (no F-F bonds)
7. XeF_5^+ (no F-F bonds)
8. SO_4^{2-} (no O-O bonds)
9. C_4H_8O
10. C_5H_{12}

List #2

1. CH_3Br (C is central atom)
2. ICl_2^+ (no Cl-Cl bonds)
3. NO_2^- (no O-O bonds)
4. BF_3 (no F-F bonds)
5. HOOH (bonding as in formula)
6. $SbCl_5$ (no Cl-Cl bonds)
7. BrF_3 (no F-F bonds)
8. ICl_4^- (no Cl-Cl bonds)
9. $CH_3CH_2CO_2^-$ (bonding as in formula)
10. C_5H_{10}

List #3

1. PO_4^{3-} (no O-O bonds)
2. H_2NNH_2 (bonding as in formula)
3. CS_2 (no S-S bonds)
4. BCl_3 (no Cl-Cl bonds)
5. SbF_5 (no F-F bonds)
6. PCl_6^- (no Cl-Cl bonds)
7. ClF_5 (no F-F bonds)
8. XeF_2 (no F-F bonds)
9. C_3H_5Cl (C's are central atoms)
10. C_6H_{12} (6 C's in a ring)

List #4

1. NH_4^+
2. SF_2 (no F-F bonds)
3. COF_2 (C is central atom)
4. SO_2 (no O-O bonds)
5. PBr_5 (no Br-Br bonds)
6. $TeCl_4$ (no Cl-Cl bonds)
7. XeF_4 (no F-F bonds)
8. ClO_4^- (no O-O bonds)
9. C_3H_8O
10. C_6H_6 (6 C's in a ring)

MOLECULAR STRUCTURES

Name_____ Section_____

Partner _____ Molecule List # _____

1.   # valence electrons = _____

Lewis Structure: Molecular geometry = _____

3-D Drawing:

2.   # valence electrons = _____

Lewis Structure: Molecular geometry = _____

3-D Drawing:

3.   # valence electrons = _____

Lewis Structure: Molecular geometry = _____

3-D Drawing:

4.   # valence electrons = _____

Lewis Structure: Molecular geometry = _____

3-D Drawing:

5.   # valence electrons = _____

Lewis Structure: Molecular geometry = _____

3-D Drawing:

6.   # valence electrons = _____

Lewis Structure: Molecular geometry = _____

3-D Drawing:

7.   # valence electrons = _____

Lewis Structure: Molecular geometry = _____

3-D Drawing:

8.   # valence electrons = _____

Lewis Structure: Molecular geometry = _____

3-D Drawing:

9.   # valence electrons = _____

Lewis Structure: Molecular geometry = _____

3-D Drawing:

10.   # valence electrons = _____

Lewis Structure: Molecular geometry = _____

3-D Drawing:

Examples of Work to be Done During the Lab Period

(Note: Only the molecular formulas will be given, you must determine the number of valence electrons, the Lewis Structure, the geometry and give a 3-D drawing.)

1. NH_3 # valence electrons = __8__

   Lewis Structure: Molecular Geometry = <u>triangular pyramid</u>

   H—N̈—H
     |
     H

   3-D Drawing:

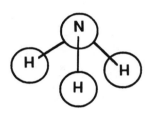

2. $CH_3CH_2CH_3$ # valence electrons = __20__

   Lewis Structure: molecular Geometry = <u>tetrahedral about each C atom</u>

```
   H   H   H
   |   |   |
H—C—C—C—H
   |   |   |
   H   H   H
```

   3-D Drawing:

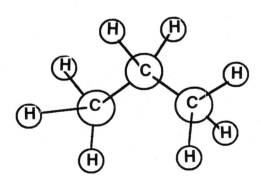

SUPPLEMENTS ON MOLECULAR STRUCTURES

The following pages contain study supplements on the various topics involved in determining molecular structures. Your textbook and the classroom lectures may also provide information on these topics.

The supplements contain activities designed so that you can interact with them. The activities are divided into sections designated by a capital letter. Most of these sections ask questions which you should answer by writing in the space provided. The sections are designed so that you can compare your answer with the correct answer before proceeding to the next section. These answers are designated by lower case letters.

Make a mask by folding a sheet of paper. Cover the sections below the one you are studying. When you have answered the question, slide the mask down to expose the correct answer and compare it with yours. If it is the same or you understand your error, proceed to the next section. If you don't understand your error, see your instructor.

The supplements are designed so that you can study (and refer back to) just those topics that are giving you difficulty.

a. Definitions and Ideas

The distinguishing characteristic of a molecule or molecular ion is that its atoms are held together by what are called *covalent bonds*. In covalent bonding two atoms are attracted to the same set of electrons. This mutual attraction to the same electrons keeps the atoms a set distance apart. The mutual attraction, in effects, bonds the atoms together. The two atoms are said to *share* the set of *bonding electrons*. The three principle types of covalent bonds are symbolized below using carbon atoms.

C—C	C = C	C ≡ C
single bond	double bond	triple bond
2 electrons shared	4 electrons shared	6 electrons shared

The only electrons that can become involved in covalent bonding are those in the outer energy shell (highest n level) of the atom. (The other electrons, those in the inner energy levels, are simply buried too deeply in the atom to be affected by an outside nucleus.) The outer shell electrons of an atom are called its *valence electrons*. In a molecule or molecular ion, some of these valence electrons will be shared in covalent bonds. Other valence electrons, called *unshared pairs* will be attracted to only one atom.

A *Lewis Structure* is a line-and-dot picture that shows where the valence electrons are and what they are doing in a molecule. The Lewis Structure for acetic acid, CH_3CO_2H, is shown below.

It shows, for example, that each hydrogen atom is held in place by a single bond and has no unshared valence electrons on it; that one oxygen has two pairs of unshared valence electrons and is sharing four others with a carbon atom (is double bonded).

The first Lewis Structures had to be determined by interpreting experimental data on the molecules. However, enough of this work has been done to enable us to *predict* Lewis Structures with reasonable certainty. In other words, the principles used by nature to distribute valence electrons in molecules have been discovered.

Controlling Ideas for Predicting Lewis Structures

There are many methods for deducing Lewis Structures, the method used in the remaining parts of this supplement may differ from the one given in your textbook. Whatever method is used, the following ideas are controlling the work.

1. **Number of Valence Electrons**. The Lewis Structure picture must show just the number of valence electrons used in the actual molecule or molecular ion.

2. **Octet (duet) Rule**. Each atom in the molecule will usually have 8 valence electrons on it (octet rule). This counts *all* the valence electrons the atom is sharing as well as its unshared pairs. Hydrogen atoms will have only 2 valence electrons (duet rule) each. (You will see that the octet rule is not always followed by the atoms in periods 3 and higher of the Periodic Table.)

Note: The method for deducing Lewis Structures given in the remainder of this supplement only works well with covalent compounds containing ''A'' Group elements; i.e., representative elements.

b. The Number of Valence Electrons

A. Generate the electron configuration for each different atom in a $HCCl_3$ molecule.

  H =

  C =

  Cl =

---

a.

$H = 1s^1$

$C = 1s^2 2s^2 2p^2$

$Cl = 1s^2 2s^2 2p^6 3s^2 3p^5$

B. The number of valence electrons an atom has can be determined in several ways. One is to generate the electron configuration of the atom and count the number of electrons in the highest n level. (These will be the outermost s and p orbital electrons.)

Determine the number of valence electrons for each atom in the molecule $HCCl_3$

  H =

  C =

  Cl =

---

b.

H = 1

C = 4

Cl = 7

C. The total number of valence electrons in a neutral molecule can be determined by summing the number of valence electrons for each atom in the molecule.

If the species is a molecular ion, one must go one step further to account for the charge. Add the extra electrons if it is an anion (negative ion) or subtract the lost electrons if it is a cation (positive ion).

Determine the total number of valence electrons in a $HCCl_3$ molecule. (Remember there are three Cl atoms.)

                    # valence electrons = _____

---

c. # val. e⁻ = 26

  1 H x 1 =   1
  1 C x 4 =   4
  3 Cl x 7 = 21
            ―――
             26

D. Thus 26 electrons must be shown in the Lewis Structure of $HCCl_3$.

For other examples of calculating the # valence electrons see the three Lewis Structure examples on the next pages.

---

c. Determining Lewis Structures

Example #1

	A. You are to determine the Lewis Structure for PCl_3, given only its molecular formula and the fact that there are no Cl-Cl bonds. What does this last fact tell you?
a. Each Cl is individually bonded to the P atom.	B. Determine the total number of valence electrons in a PCl_3 molecule. # valence electrons = _____
b. # val. e⁻ = 26 1 P x 5 = 5 3 Cl x 7 = 21 26	C. We will ignore this number for the moment and just guess at a possible Lewis Structure. Recalling answer a, connect the appropriate atoms with single bonds.

c.
 	
Cl—P—Cl
     |
    Cl
	D. Now add unshared electron *pairs* to each atom until each has 8 (shared plus unshared) electrons on it. If hydrogen atoms were present, each would only have 2 electrons.

d.
:C̈l—P̈—C̈l:
    |
   :C̈l:
	E. This is our "first guess" at the Lewis Structure and must be checked for number of electrons. How many electrons are shown in the picture?

e. 26 e⁻ 10 unshared pairs 3 bonded pairs = 13 pairs = 26 e⁻	F. This number is exactly what we calculated that nature uses in the molecule (# valence electrons). Since our "first guess" Lewis Structure contains the correct number of electrons and since each atom has an octet, this is a correct Lewis Structure for PCl_3.

Example #2

	A. You are to determine the Lewis Structure for the molecular ion CO_3^{2-} (no O-O bonds). Determine the total number of valence electrons in one CO_3^{2-} ion. # valence electrons = _____

a. # val. e⁻ = 24
 	
   1 C x 4 = 4
 	
   3 O x 6 = 18
 	
   extra e⁻ = _2_
 	
                24
	B. The species is charged because it has 2 more electrons than those contributed by the neutral atoms.
	
	Recalling that there are no O-O bonds, draw a picture connecting the appropriate atoms with single bonds.

b. O—C—O
         |
         O
	C. Now add unshared electron pairs until each non-hydrogen atom has a total of 8.

c. :Ö—C̈—Ö:
         |
       :Ö:
	D. Count the number of electrons in this "first guess" picture.
	
	How does this number compare to the calculated # val. e⁻ (ans. a)?

d. 26 e⁻ shown.
   This is 2 e⁻ too
   many, since 24 e⁻
   are required.
	E. Two electrons must be removed from our "first guess" picture. Remove them as an unshared pair from the central atom; i.e., the carbon atom.

117

e. :Ö—C—Ö:
 	
     |
    :O:
	F. This structure has the required 24 electrons but now the C atom has only 6 electrons instead of the required 8.
	
	This problem is solved by turning an unshared pair on one of the oxygens into a bonded pair. That is, by using an unshared pair to create a double bond,
	
	i.e., —C—Ö:
	       |
	
	Redraw the picture using this idea.

f. :Ö—C=O:
 	
     |
    :O:
	G. This Lewis Structure contains the required 24 electrons and each atom has an octet. Thus it is a correct Lewis Structure.

In general, whenever your "first guess" picture contains too many electrons, the correct Lewis Structure will contain double or triple bonds. (These so called multiple bonds occur most often between atoms of C, N, O, P, S, Se. These six multiple bonders are easy to recall because they form a triangle in the periodic table.)

A molecule where the multiple bond can be located in two or more positions is best represented by a series of Lewis Structures that is given the term *resonance*. Consult your text or your instructor for details.

<u>Example # 3</u>

	A. You are to determine the Lewis Structure for the molecule SF_4 (no F-F bonds). Determine the total number of valence electrons in one SF_4 molecule. # valence electrons = _____

a. # val. e^- = 34
 	
  1 S x 6 = 6
 	
  4 F x 7 = <u>28</u>
           34
	B. Draw a picture connecting the appropriate atoms with single bonds. Recall that SF_4 has no F-F bonds.

b.   F
 	
             |
   F—S—F
            |
            F
	C. Now add unshared electron pairs until each non-hydrogen atom has a total of 8 around it.

c.   :F̈:
 	
        |
   :F̈—S—F̈:
        |
        :F̈:
	D. Count the number of electrons in this "first guess" picture.
	
	How does this number compare to the calculated # val. e⁻ (ans. a)?

d. 32 e⁻ shown.
   This is two too few,
   since 34 are required.
	E. Two electrons must be added to our "first guess" picture. There is no way to do this without giving some atom more than 8 electrons. However, this is what nature is doing, the octet rule is not upheld.
	
	Place the two electrons, in the form of an unshared pair, on the central atom, i.e., the sulfur atom.

e. :F̈╲   ╱F̈:
       S̈
  :F̈╱   ╲F̈:
	F. This Lewis Structure contains the required 34 electrons. Although the octet rule is not upheld, this is so for only one atom, the central atom. Thus it is a correct Lewis Structure.

In general whenever your "first guess" picture contains too few electrons, the octet rule will be broken on the central atom. (The central atom will be in the third or higher period of the Table. Such atoms have d orbitals of low enough energy to take the "extra" electron pairs.)

Note that multiple bonds cannot help in such "too few electrons" cases.

Supl. #2—GEOMETRIES OF MOLECULES

a. Definitions and Ideas

This page contains all the essential ideas needed to determine the geometry of molecules. However, few students will be able to understand these ideas from this page alone. Understanding will come from doing the exercises in the remainder of this supplement. Use this page to get some initial feeling for the ideas now. Use it later as a quick reference on definitions.

The Lewis Structure of a molecule indicates which atoms are bonded to which and where all the valence electrons are in the molecule. A Lewis Structure does not show the geometric arrangement of the atoms in a molecule. However the Lewis Structure is the starting point for determining molecular geometries.

The ideas used to determine geometries from Lewis Structures are called the *Valence-Shell Electron-Pair Repulsion Theory* (abbreviated VSEPR theory). The ideas are as follows. The sets of valence electrons on each atom are free to move around, but not away from, the atom (much like a dog on a short leash can move around his handler). Since each set of valence electrons is negatively charged, the individual sets repel each other. This repulsion forces the individual sets to place themselves around the atom so that they are as far from each other as possible. Thus the sets of electrons occupy a fixed geometry around the atom. Any other atoms sharing the electron sets (i.e., bonded atoms) will also occupy this fixed geometry.

The individual sets of valence electrons that can move away from each other are often called *independent regions of electron density*. For geometry purposes each of the following is considered as one independent region of electron density:

    .., each unshared pair is 1 region
    __, each single bond is 1 region
    =, each double bond is 1 region
    ≡, each triple bond is 1 region

The geometry of a molecule is determined from the number of such regions on each centrally located atom. A *central atom* is one which has two or more other atoms bonded to it. A given number of regions on a central atom will assume the fixed geometry given in the table on page 127. This fixed geometry of the regions is sometimes called the *electron geometry*. Bonded atoms "tag along" with the electron regions of the central atom, taking the same position in space as the electron region they are sharing. The *molecular geometry* describes the positions the bonded atoms take around the central atom. If the central atom has all of its electron regions used in bonding, the electron and molecular geometries are identical. If the central atom has unshared pairs, the molecular geometry is a "daughter" of the electron geometry as shown in the last column in the table on page 127.

A. The determination of molecular geometries is based on the idea that independent regions of electron density (negative charge) repel each other while they are simultaneously attracted to regions of positive charge. Consequently, these regions of electron density will arrange themselves to be as far apart as possible while they are simultaneously attracted to a nucleus.

B. If an atom has two independent regions of electron density ( ) arranged about a nucleus ( ) which of the following represents the best arrangement?

b.

C. In this arrangement they are simultaneously close to the positive charge and far from each other. This arrangement is called *linear*. Show how three independent regions of electron density arrange themselves around a nucleus.

c.

D. This arrangement is called a *triangular plane*. The three areas of electron density are arranged so that imaginary lines drawn between them define an equilateral triangle which is in a plane (i.e., the triangle is in two dimensions).

E. If an atom has 4, 5, or 6 independent regions of electron density arranged around its nucleus, the regions achieve maximum separation with three dimensional arrangements. These are called tetrahedral, triangular bipyrimidal and octahedral arrangements, respectively. These arrangements are pictured on page 127 and models of them will be displayed in the lab. Study these until you can properly visualize them in your own mind.

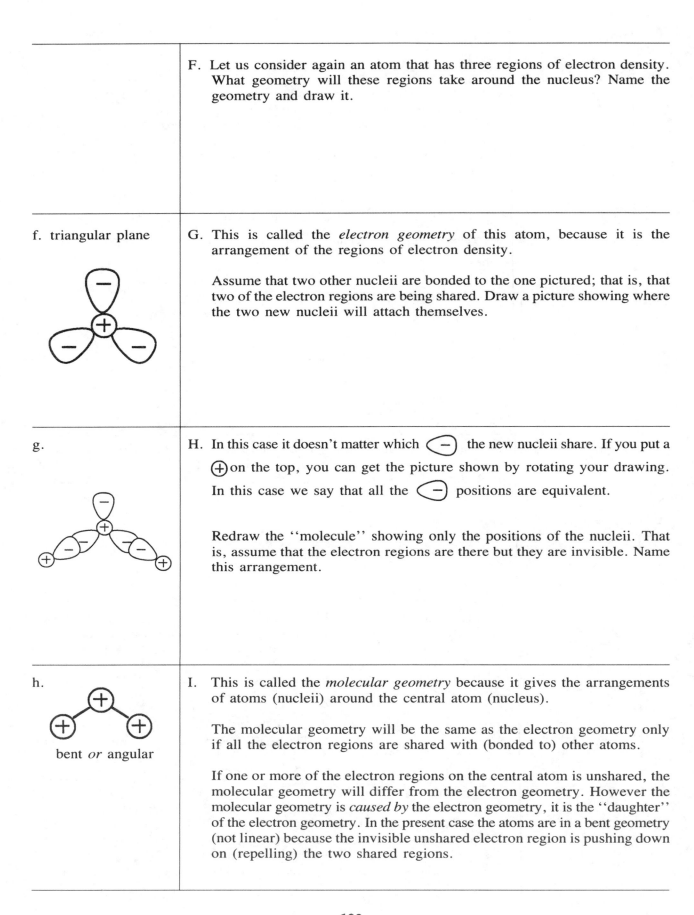

F. Let us consider again an atom that has three regions of electron density. What geometry will these regions take around the nucleus? Name the geometry and draw it.

f. triangular plane

G. This is called the *electron geometry* of this atom, because it is the arrangement of the regions of electron density.

Assume that two other nucleii are bonded to the one pictured; that is, that two of the electron regions are being shared. Draw a picture showing where the two new nucleii will attach themselves.

g.

H. In this case it doesn't matter which ⊖ the new nucleii share. If you put a ⊕ on the top, you can get the picture shown by rotating your drawing. In this case we say that all the ⊖ positions are equivalent.

Redraw the ''molecule'' showing only the positions of the nucleii. That is, assume that the electron regions are there but they are invisible. Name this arrangement.

h.

bent *or* angular

I. This is called the *molecular geometry* because it gives the arrangements of atoms (nucleii) around the central atom (nucleus).

The molecular geometry will be the same as the electron geometry only if all the electron regions are shared with (bonded to) other atoms.

If one or more of the electron regions on the central atom is unshared, the molecular geometry will differ from the electron geometry. However the molecular geometry is *caused by* the electron geometry, it is the ''daughter'' of the electron geometry. In the present case the atoms are in a bent geometry (not linear) because the invisible unshared electron region is pushing down on (repelling) the two shared regions.

c. Determining Molecular Geometries

Example # 1

	A. You are to determine the molecular geometry of CO_3^{2-} (no O-O bonds). Determine the Lewis Structure of the molecular ion.
a. $:\ddot{O}-C=\ddot{O}$ $\quad\quad :\!\underset{\cdot\cdot}{\overset{|}{O}}\!:$	B. Which is the central atom in the molecule? $\quad\quad\quad\quad\quad$ central atom = _____
b. The C atom	C. How many independent regions of electron density are on this central atom? $\quad\quad\quad\quad\quad$ # regions = _____
c. 3 regions	D. Note that the double bond counts as only one region because the 4 electrons in the double bond must stay together. How will the 3 regions arrange themselves around the carbon? That is, what is the electron geometry? (Consult the Table on page 127.) $\quad\quad\quad\quad\quad$ electron geometry = _____
d. triangular plane	E. How many of the three regions are used in bonding to other atoms? $\quad\quad\quad\quad\quad$ # bonded regions = _____
e. all 3	F. What will be the arrangement of oxygen atoms around the carbon? What is the molecular geometry? $\quad\quad\quad\quad\quad$ # molecular geometry = _____
f. triangular plane 	G. The molecular geometry is identical to the electron geometry because all the electron regions on the central atom are shared (bonding) electron regions.

Example # 2

A. You are to determine the molecular geometry of SF_4 (no F-F bonds).

Determine the Lewis Structure of the molecule.

---

a.

B. Which is the central atom in the molecule?

central atom = _____

---

b. the S atom

C. How many independent regions of electron density are on this central atom?

regions _____

---

c. 5 regions.
   1 unshared pair
   + 4 single bonds

D. According to the table on page 127, how will these 5 regions arrange themselves around the S atom?

electron geometry = _____

---

d. triangular
   bipyramid

E. To properly see this geometry you should inspect the model on display in the lab or build your own with the materials provided. The geometry has 3 regions in a triangular plane (t regions). The other two are directly above and below the triangle, forming the apex (a regions) of two pyramids.

How many of the 5 regions are used in bonding?

bonded regions = _____

---

e. 4

F. Since not all regions are used in bonding, the molecular geometry will not be the same as the electron geometry but will be some derivative of it. We must determine where the unshared pair will be located in the electron geometry.

Refer to the figure in E above. If the unshared pair locates in one of the apex (a) positions, the molecular geometry would be a (single) triangular pyramid. If the unshared pair locates in one of the triangle (t) positions, a seesaw molecular geometry is obtained. (The other two t positions would be the legs; the a positions, the seats of the seesaw.)

	G. According to the last column in the table on page 127, which of the possible geometries is actually used in nature? molecular geometry = _____
g. seesaw	

Example #3

	A. You are to determine the molecular geometry of a methylamine molecule, CH_3NH_2. Determine the Lewis Structure of the molecule (the H atom bonding is indicated by the formula).

a.
 	
        H
        |
H—C—N̈—H
        | |
       H H
	B. Which is (are) the central atom(s) in the molecule?
	
	central atom(s) = _____

b. both the C and the N atoms	C. In cases where there is more than one central atom, the total geometry depends on the geometry around each central atom. We will apply the information in the table on page 127 to each central atom individually and put the two geometries together only in a final step. How many independent regions of electron density are on each central atom? # regions: C = _____ N = _____
c. C = 4 N = 4	D. Note that the bond between the C and the N is counted twice, as a region belonging to each atom. How will the regions around each central atom arrange themselves? electron geometries: C = _____ N = _____

d. C = tetrahedron
    N = tetrahedron
	E. How many of the regions around each central atom are used in bonding?
	
	    # bonded regions: C = _____
	
	    N = _____

e. C = 4
    N = 3
	F. According to the Table what will be the arrangement of atoms around each central atom?
	
	    molecular geometries: C = _____
	
	    N = _____

f. C = tetrahedron
    N = triangular
        pyramid
 	
    triangular pyramid
 	
    tetrahedron
	G. We are now ready to put these individual geometries together to form the molecule.
	
	    Naming the total molecular geometry is best done by just stating the individual geometries around each central atom; i.e., by giving answer f.
	
	    Making a model of the molecule is a bit more involved. The molecule is a tetrahedron sharing an atom with a triangular pyramid. Make separately a tetrahedron and a triangular pyramid. Now remove a "bond" stick from one model and insert a "bond" stick from the other model in its place. The result is illustrated.

Table of Geometries

Number of regions of electron density	Arrangement of regions for maximum separation	Number of regions used in bonding	Geometry of molecule
2	linear	2	linear
		1	linear
3	triangular plane	3	triangular plane
		2	angular
		1	linear
4	tetrahedron	4	tetrahedron
		3	triangular pyramid
		2	angular
		1	linear
5	triangular bipyramid	5	triangular bipyramid
		4	sea saw
		3	T
		2	linear
6	octahedron	6	octahedron
		5	square pyramid
		4	square plane

Often it is necessary to try to represent three dimensional molecular geometries in drawings. The following instructions are intended to illustrate a method for drawing 3-D geometries. The important consideration is to obtain realistic drawings which represent the geometry of the molecule.

A. In order to keep the situation simple, atoms in a molecule will be represented by circles. Smaller atoms can be represented by smaller circles. The bond attaching atoms will be represented by lines connecting the atoms. For example, the molecule HCl can be represented as:

B. Molecules with geometries which are in one plane can be represented on paper quite easily. The difficulty arises with some of the molecules which are based on the tetrahedron, triangular bipyramid and octahedron. Some of these are 3-D and therefore, require perspective drawings.

C. A technique for suggesting perspective is to show the bonds overlapping the circle which is behind the plane of the paper.

Fig. 1   Fig. 2   Fig. 3

In figure 1 both Cl and H are in the plane of the paper. In figure 2 the Cl is behind the H and in figure 3 Cl is in front of H.

D. Consider the molecule CH_4 which has a Lewis Structure of $\quad H-\overset{\displaystyle H}{\underset{\displaystyle H}{C}}-H.\quad$ Since there are 4 bonded electron pairs around the C atom and no non-bonded pairs, the geometry around the C atom must be tetrahedral. A 3-D drawing of this geometry is obtained as follows. (See the table on page 127.)

E. Assume you have made a 3-D ball and stick model for CH_4. You rotate it in your hand until the central atom and one of the hydrogens are in one plane. A drawing of that might look like this:

F. If this orientation is rotated until two other hydrogens are in front of the plane and one is behind, the placement of circles might look like this:

G. Then if lines were drawn so that the two in front overlap the carbon and the one in back overlaps one hydrogen, the final drawing looks like this:

H. The following steps are suggested for deriving 3-D drawings:

(1) Build a 3-D model of the molecule.
(2) Rotate the model until as many atoms as possible are in the same plane as the central atom. Make sure all the atoms are showing.
(3) Draw circles representing the placement of all atoms in the molecule.
(4) By considering the central atom to be in the plane, draw all the atoms as being in the same plane, in front of, or behind the central atom by overlapping bond lines.

STUDIES OF A PRECIPITATION REACTION—Exp. I-1

Name_____ Lab Section_____

Lab Partner _____

Pre-lab assignment: Study Appendix D.1 on volume measuring and study Appendix D.5 on filtration techniques.

I. The Interaction of Calcium Nitrate and Potassium Iodate

    A. Obtain, in clean labeled beakers, about 80 mL of $Ca(NO_3)_2$ and 80 mL of KIO_3 solution. Record the molar concentrations of these stock solutions in the table on page 132.

    Using volumetric cylinders or pipets with bulbs prepare the following solutions in clean labeled beakers.

    Solution 1: 25.0 mL $Ca(NO_3)_2$ + 25.0 mL KIO_3

    Solution 2: 15.0 mL H_2O + 10.0 mL $Ca(NO_3)_2$ + 25.0 mL KIO_3

    Mix each solution with a separate stirring rod rubbing the tip of the rod against the bottom or sides of the beakers (This is called scratching the solution.). Record your observations below.

    Let these solutions stand for 15-20 minutes, stirring them occasionally. (You will want to recover all the precipitate later so it is best to leave the stirring rods in the solutions.) In the meantime proceed with part I.B.

    B. Assuming the reaction involves the coming together of dissolved ions, what are the possible identities of the precipitate formed? Think up and carry out experiments which will distinguish between the possibilities. Describe the results of these experiments.

    C. Write a balanced chemical equation that represents the reaction and is consistent with the data obtained so far.

II. Weight of the Precipitates

A. Label and weigh two pieces of filter paper. Record these masses in the table below. Following the directions in Appendix D.5, filter each solution catching the supernatants in labeled clean beakers.

Using the appropriate supernatants wash onto the filter paper any precipitate that remains in the beaker or on the rod. When all the precipitates have been trapped in the paper, set the supernatant solutions aside for later use.

Rinse the precipitates with four separate 5 mL portions of methanol. (**CAUTION**: the methanol is flammable—use no open flames.) Gently remove the filter papers from the funnels and place them on watch glasses. Let these air dry for about 5 minutes then place them in a 90-110°C drying oven. Leave them in the oven to dry for at least 30 minutes, then cool, weigh and record the masses in the table. In the meantime proceed with the remaining parts of the experiment.

Concentration of Stock $Ca(NO_3)_2$ _____M

Concentration of Stock KIO_3 _____M

	Solution #1	Solution #2
a. mL $Ca(NO_3)_2$ solution	_____	_____
b. mL KIO_3 solution	_____	_____
c. mass of filter paper	_____	_____
d. mass of paper & ppt.	_____	_____
e. mass of ppt.	_____	_____
f. moles of $Ca(NO_3)_2$	_____	_____
g. moles of KIO_3	_____	_____
h. expected mass of ppt.	_____	_____

B. Calculate the number of moles of $Ca(NO_3)_2$ and KIO_3 added to each solution. Using these amounts and the chemical equation you deduced in part I.C, calculate the expected mass of the precipitates (is there a limiting reagent in each case?) Show your calculations and reasoning below and record the numbers in parts f, g, h of the table.

III. Tests of the Supernatants

A. Obtain about 5 mL of 0.3 M Na_3PO_4, 5 mL of 1 M HCl and about 2 cm³ of solid KI. Place about 1 mL of the stock $Ca(NO_3)_2$ and KIO_3 solutions in separate test tubes.

Ca^{2+} test: add about 10 drops (1/2 mL) of Na_3PO_4 solution to the 1 mL of $Ca(NO)_2$ solution. Describe the results.

(The reaction is: $3Ca^{2+}(aq) + 2PO_4^{3-}(aq) \longrightarrow Ca_3(PO_4)_2(s)$.)

IO_3^- test: dissolve about 1/2 cm³ of KI into the 1 mL of KIO_3 solution then add about 1 mL of 1 M HCl. Describe the results.

(The reaction is $IO_3^-(aq) + 5I^-(aq) + 6H^+(aq) \longrightarrow 3I_2(aq) + 3H_2O$.)

B. Using these tests check the two supernatants from part II.A for the presence of Ca^{2+} and IO_3^-. Describe the results.

IV. Dropwise Mixing

A. Using a dropper add the stock $Ca(NO_3)_2$ solution 5 to 10 drops at a time to your remaining sample of stock KIO_3 solution. After each addition stir with scratching (rubbing the stirring rod against the beaker bottom or sides). Record your observations after each addition.

V. Analysis of Results

A. How do the actual precipitate masses compare to the masses you predicted by calculation? If they do not compare well make a list of at least 4 possible explanations.

B. Are the test results for the presence of Ca^{2+} and IO_3^- in the supernatants consistent with the limiting reagent, excess reagent calculations in part II.B? Are there any inconsistencies between the calculated and test results? Explain your reasoning.

C. Were the results of the Dropwise Mixing experiment unexpected? (i.e. Given the chemical equation deduced in part I.C, what should have happened with the first addition of $Ca(NO_3)_2$ to KIO_3?) Explain your reasoning.

D. Offer a single reason that will explain the results in all three experiments. (Hint: will any of the possible explanations you used in part V.A also work for the results of the other two experiments?) Show how your single explanation fits all the data collected.

PROPERTIES OF CALCIUM IODATE—Exp. I-2

Name_____ Lab Section_____

Lab Partner _____

Pre-lab assignment: Study Appendix D.1 on volumetric measurements and titrations.

I. Preparation of Solutions

To save time one partner should work on part A below while the other partner works on part B. **NOTE**: There are two stock solutions of both $Ca(NO_3)_2$ and KIO_3 available. Be sure to check the concentrations to insure that you are using the proper solution.

A. Preparations of $Ca(IO_3)_2(s)$

Obtain 20 mL of 23 g/100 mL $Ca(NO_3)_2$ solution and 70 mL of 5 g/100 mL KIO_3 solution in separate clean beakers. Pour the solutions together and stir. Rub the stirring rod against the sides or bottom of the beaker. (This is called scratching the solution and is used to speed the formation of the precipitate.)

The precipitate formed is $Ca(IO_3)_2$. Let this solution stand for 10 minutes and then filter (See Appendix D.5). Using the supernatant wash any precipitate that remains in the beaker onto the filter paper until about 90% of the solid has been trapped on the paper. Wash the trapped precipitate with four separate 10 mL portions of distilled water.

B. Preparation of $Ca(NO_3)_2$ Solutions.

Obtain 90 mL of 0.1 M $Ca(NO_3)_2$ in a clean, dry beaker. Record its exact concentration in the space below. 40 mL of this solution will be used in part C, the remainder will be used to make less concentrated solutions of about 0.025 M and 0.006 M.

Rinse a 25 mL pipet with a 5 mL portion of the 0.1 M solution. Using a pipetting bulb (See Appendix D.1) transfer 25.00 mL of the 0.1 M solution into a 100.00 mL volumetric flask. Fill to the mark with distilled water and mix by inversion. Label a clean beaker and rinse it with two 5 mL portions of the solution. Store the solution in the beaker.

Rinse the pipet and volumetric flask with distilled water and then prepare a 0.006 M solution in a similar manner using the 0.025 M solution. Calculate the concentrations of these two prepared solutions to 3 significant figures in the following space.

C. Label four clean 100 mL beakers as # 1 to # 4. Put 40 mL of a different solution (distilled water, 0.006 M Ca^{2+}, 0.025 M Ca^{2+}, 0.1 M Ca^{2+}) into different beakers. Each beaker should be rinsed with a small portion of its solution before the 40 mL is added. Record which beaker contains which solution in table 2 on page 141.

Put about 1/4 of the wet $Ca(IO_3)_2$ precipitate into each beaker. Stir each mixture (use a separate stirring rod for each to prevent contamination). Let the solutions stand for about 30 minutes with occasional stirring. Record any observations below. In the meantime proceed with part II.

II. Standardization of Sodium Thiosulfate Solution

A. Obtain 100 mL of sodium thiosulfate ($Na_2S_2O_3$) solution in a clean, labeled beaker. Rinse a buret with small portions of the solution and fill the buret with the solution.

B. Obtain about 50 mL of 0.01 M KIO_3 solution, 10 cm³ of solid KI, 20 mL of 1 M HCl and 20 mL of 0.2% starch solution. Obtain a medicine dropper for the HCl solution and one for the starch solution. Using a bulb rinse the 10 mL pipet with three 2 mL portions of the 0.01 M KIO_3. Pipet 10.00 mL of this solution into each of three distilled water rinsed 250 mL Erlenmeyer flasks. Add 10 to 20 mL of distilled water. Dissolve about 1 cm³ of solid KI into each flask and add two droppersful of 1 M HCl (approximately 2 mL). Swirl to mix.

The coloring of the solution comes from the I_2 formed by the oxidation-reduction reaction:

$$IO_3^-(aq) + 5\ I^-(aq) + 6H^+(aq) \longrightarrow 3I_2(aq) + 3H_2O$$

C. Record the initial buret volume reading of thiosulfate solution in table 1 on page 139. Select one I_2 solution flask and titrate it with the $S_2O_3^{2-}$ solution (See Appendix D.1 for pointers). Be sure to swirl the flask after each **small** addition of $S_2O_3^{2-}$. The end point of the titration is reached at just that point where the titrated solution is permanently colorless. It is at this point that the colored I_2 has just been consumed by the $S_2O_3^{2-}$. The reaction is:

$$2\ S_2O_3^{2-}(aq) + I_2\ (aq) \longrightarrow S_4O_6^{2-}(aq) + 2I^-(aq)$$

The color of the solution will fade from yellow-brown to pale yellow as the titration progresses. When the pale yellow color is reached, add a dropperful of starch solution. The starch simply enhances the coloring of I_2 making it easier to determine the end point. Continue the titration until the colorless end-point is reached. Record the final buret reading and the total volume of $S_2O_3^{2-}$ solution used.

Titrate the other I_2 solutions. If the results of the three titrations do not agree to within 1%, run a fourth titration.

D. Calculate the molar concentration of the stock $S_2O_3^{2-}$ solution from the known concentration of your IO_3^- solution and the combination reaction (See B and C above):

$$IO_3^-(aq) + 6 H^+(aq) + 6 S_2O_3^{2-}(aq) \longrightarrow I^-(aq) + 3 S_4O_6^{2-}(aq) + 3 H_2O$$

Show your calculations for the second titration in the space below. Calculate an average value for $S_2O_3^{2-}$ solution concentration.

TABLE 1
Standardization of Stock Thiosulfate Solution

Concentration of KIO_3 solution titrated _____ M

titration no.	1	2	3	4
mL KIO_3	_____	_____	_____	_____
initial buret reading (mL)	_____	_____	_____	_____
final buret reading (mL)	_____	_____	_____	_____
total $S_2O_3^{2-}$ volume (mL)	_____	_____	_____	_____
molarity of $S_2O_3^{2-}$ (M)	_____	_____	_____	_____
average $S_2O_3^{2-}$ molarity	_____ M			

III. Analysis of the Working Solutions

A. You are to filter each solution prepared in part I.C so that a 20 to 40 mL sample of uncontaminated and undiluted supernatant is obtained. Label four clean and dry beakers. Prepare a dry filter paper cone for each of four clean and dry filter funnels. Pass each solution through a different dry paper cone catching the supernatant in the appropriately labeled beaker. The solid $Cu(IO_3)_2$ and filter papers can be discarded.

B. Label four distilled water rinsed Erlenmeyer flasks and pipet 10.00 mL of each supernatant into a different flask.

Add 10-20 mL of distilled water, 1 cm^3 of solid KI and two droppersful of 1 M HCl to each flask. Record your observations.

C. What conclusions can be drawn from these observations? (Hint: what ions must be present for the colored I_2 to form, where did each of these come from in these experiments? Considering the data collected to this point, what can be said about the extent to which $Ca(IO_3)_2$ dissolves?)

D. Titrate each of these solutions as was done in part II. Record the titration data in table 2. If time permits titrate a second 10.00 mL sample of each supernatant.

Using the molar concentration of stock thiosulfate solution determined in part II and the analysis reaction given there, determine the molar concentration of IO_3^- developed in each supernatant solution. Record these values in table 2. Show your calculations on supernatant # 2 in the space below.

TABLE 2

Analysis of Working Solutions

Solution No.	1	2	3	4
initial $[Ca^{2+}]$, (M)	_____	_____	_____	_____
initial buret reading (mL)	_____	_____	_____	_____
final buret reading (mL)	_____	_____	_____	_____
total $S_2O_3^{2-}$ volume (mL)	_____	_____	_____	_____
molarity IO_3^- (M)	_____	_____	_____	_____

IV.A. Transfer from table 2 to table 3 the initial concentration of Ca^{2+} and the final concentration of IO_3^- for each supernatant. Assuming that $Ca(IO_3)_2$ completely breaks into IO_3^- and Ca^{2+} ions when it dissolves, calculate the concentration of $Ca(IO_3)_2$ that dissolved and the final concentration of Ca^{2+}. Show the calculations for the second supernatant below.

TABLE 3
Supernatant Compositions (all concentrations in M)

Solution No.	1	2	3	4
initial $[Ca^{2+}]$	_____	_____	_____	_____
final $[IO_3^-]$	_____	_____	_____	_____
$[Ca(IO_3)_2]$ dissolved	_____	_____	_____	_____
final $[Ca^{2+}]$	_____	_____	_____	_____

B. Identify, qualitatively, any patterns that exist in the data in table 3. Offer a physical explanation for these patterns. (Hint: write a chemical equation to represent the dissolving of $Ca(IO_3)_2(s)$. How does the initial presence of $Ca^{2+}(aq)$ affect the dissolving? Why might this be?)

C. Find an algebraic expression that best relates the final concentrations of IO_3^- and Ca^{2+} to each other. (Hint: try all possible mathematical combinations of the concentrations for a given supernatant using multiplication and division and squaring and cubing the concentrations. See which combination gives the most constant result for all supernatants.) What practical uses could your algebraic expression have?

ACETIC ACID—Exp. I-3

Name_____ Lab Section_____

Lab Partner _____

Pre-lab assignment: In this laboratory you will be measuring the concentration of H^+ ion (sometimes symbolized H_3O^+) by taking measurements of the pH of solutions. The pH is related to the H^+ concentration according to the equation:

$$pH = -\log[H^+]$$

The mathematical conversion of pH to $[H^+]$ and vice versa is discussed in Appendix D.3 and in your textbook. These are to be studied before doing the lab work.

You are also to study Appendix D.3 on the use of a pH meter and Appendix D.1 on the proper use of pipets and volumetric flasks.

CAUTION!!!!!

In these laboratories we will be investigating and using some of the properties of acids and bases. Due to the possible injurious effects of acids and bases on both metallic equipment and the human body, the following precautions should be followed at all times:

1. Never allow the solid acids and bases or their solutions to come into contact with your person or clothing. If such contact does occur wash the area profusely with cold water. Wear your safety glasses at all times.
2. If a solid or solution should spill, wipe up the spill immediately. Rinse the area a couple times with water and wipe dry.
3. Do not set chemicals directly on the analytical balance pans, use preweighed beakers or flasks to hold the chemicals.
4. If any large spills occur, notify the instructor immediately.

I. A. Prepare about 100 mL of an approximately 0.1 M solution of each of the acids listed below. The concentration of the stock solutions is indicated.

Acid	Concentrations of Stock Solutions	Amount Used to Prepare 0.1 M Solution
Hydrochloric Acid, HCl	4 M	
Nitric Acid, HNO_3	4 M	
Sulfuric Acid, H_2SO_4	4 M	
Acetic Acid, $HC_2H_3O_2$	4 M	
Oxalic Acid, $H_2C_2O_4$ $(H_2O)_2$		
Potassium Acid Phthalate $HKC_6H_4(CO_2)_2$		

B. Measure the pH of each of the 0.1 M solutions using pH indicator paper.

Acid	pH	$[H^+]$
HCl	_____	_____
HNO_3	_____	_____
H_2SO_4	_____	_____
$HC_2H_3O_2$	_____	_____
$H_2C_2O_4$	_____	_____
$HKC_6H_4(CO_2)_2$	_____	_____

Calculate the $[H^+]$ corresponding to these pH's and record them in the table. Show one such calculation below.

C. What conclusions can be drawn from these data? (Hint: write a chemical equation expressing the reaction that occurs when the acids are dissolved in water. Considering that all the solutions were about 0.1 M, what would the expected pH be for each solution? Which solutions attained this value and which didn't? What might be the reason for some solutions not attaining this value?)

II. A. In the laboratory you will find a stock acetic acid solution with a total concentration of 1.00 M. Use this solution to prepare solutions whose total concentrations are: 0.100 M, 0.0100 M, 0.00200 M, and 0.00100 M according to the instructions in B. (If the stock solution is not 1.00 M, record the exact concentration in the table on page 147 and prepare solutions that are 1/10, 1/100, 2/1000 and 1/1000 as concentrated as the stock solution. Record the exact total concentrations of these solutions in the table.)

B. Using a clean towel-dried beaker, obtain 60 mL of the stock $HC_2H_3O_2$ solution. Rinse a 10 mL pipet and a 100 mL volumetric flask with distilled water. Rinse the pipet with about 5 mL of the 1.00 M solution. Pipet exactly 10.0 mL of the solution into the volumetric flask. Add distilled water to bring the meniscus to the mark on the neck of the flask. Stopper the flask and mix by repeatedly inverting the flask. Pour this 0.10 M solution into a labeled, clean towel-dried beaker. Using the 0.1 M solution and a procedure similar to the above, prepare the 0.010 M solution. This solution can then be used to prepare the 0.0020 M and 0.0010 M solutions.

C. Measure the pH of each of these solutions by using a pH meter. Record the data in the table on page 147. (The pH meter and its electrodes are a very delicate and expensive set of equipment. Instructions for its use are provided by your instructor and/or are found in Appendix D.3 **FOLLOW THESE INSTRUCTIONS TO THE LETTER**. Check with your laboratory instructor if you have any questions.) Save the 0.100 M solution for part III.

D. Assuming the reaction is:

$$HC_2H_3O_2 \rightleftharpoons H^+ + C_2H_3O_2^-$$

acetic   hydrogen   acetate

acid   ion   ion

use the measured pH and the total concentration to calculate the actual concentration of each species, show the calculations for one solution below.

Conc. of Stock Solution _____

Total Concentration	pH	$[H^+]$	$[C_2H_3O_2^-]$	$[HC_2H_3O_2]$
0.100 M	_____	_____	_____	_____
0.0100 M	_____	_____	_____	_____
0.00200 M	_____	_____	_____	_____
0.00100 M	_____	_____	_____	_____

E. What conclusions can be drawn from this data? Find the algebraic equation which best relates $[H^+]$, $[C_2H_3O_2^-]$, and $[HC_2H_3O_2]$ to each other. (Hint: Try all the possible combinations of the three concentrations by multiplication and/or division. For example multiply all these together, multiply two and divide by the third, etc. looking for the combination which gives the most constant result for all sets of data.) Summarize your findings below.

147

III. A. Weigh out approximately 0.5 gram and 1.5 gram samples of anhydrous sodium acetate ($NaC_2H_3O_2$). Record the exact masses and calculate the moles in each sample.

	Grams	Moles
"0.5 g" sample	_____	_____
"1.5 g" sample	_____	_____

Using a graduated cylinder, measure exactly 50.0 mL of the 0.10 M $HC_2H_3O_2$ solution into a clean towel-dried 100 mL beaker. Measure its pH. Dissolve the "0.5 g" sample of $NaC_2H_3O_2$ in this solution and measure its pH. Then dissolve the "1.5 g" sample in the same solution and measure its pH. Calculate the total concentrations of added $NaC_2H_3O_2$ and record these in the table. Considering the measured pH values and the fact that $NaC_2H_3O_2$ completely dissociates in water to give Na^+ and $C_2H_3O_2^-$, calculate the actual concentrations of H^+, $C_2H_3O_2^-$ and $HC_2H_3O_2$ in each solution. Show your calculations for the more concentrated solution below (i.e., the 0.5 g plus 1.5 g solution).

Total Conc. of Acid	Total Conc. of $NaC_2H_3O_2$	pH	Actual Concentrations [H^+]	[$C_2H_3O_2^-$]	[$HC_2H_3O_2$]
.100M	0	_____	_____	_____	_____
.100M	_____	_____	_____	_____	_____
.100M	_____	_____	_____	_____	_____

B. Explain qualitatively (in non-mathematical terms) why the pH varies in these three solutions.

C. What quantitative (mathematical) conclusions can be drawn from this data? (Hint: does the data fit the algebraic equation you developed in part II.E? Explain your findings.)

PROPERTIES OF TWO REACTIONS—Exp. I-4

Name_____ Lab Section_____

Lab Partner _____

I. Fe(NO₃)₃ + KSCN

    A. Put about 50 mL of KSCN (potassium thiocyanate solution labeled 0.0005 M) in a 100 mL beaker. Note the appearance of the solution. Add about 5 drops of $Fe(NO_3)_3$ (Iron (III) nitrate solution labeled 0.2 M). Note the appearance of the $Fe(NO_3)_3$ solution, the KSCN solution, and the combined solution. What evidence is there for a chemical reaction between $Fe(NO_3)_3$ and KSCN?

        The observed reaction is: $Fe^{3+}(aq) + SCN^-(aq) \longrightarrow FeSCN^{2+}(aq)$. Identify the color of each reactant and product ion.

    B. Fill each of three test tubes about 1/3 full with the solution made in part I.A above. Keeping one of the test tubes as a standard of comparison, put a few crystals of KSCN into one of the other test tubes and one drop of $Fe(NO_3)_3$ solution into still another. Note observations below. How can you account for these observations (i.e., what is the meaning of any changes noted when KSCN was added? When $Fe(NO_3)_3$ was added?).

C. Fill a test tube 1/2 full of the solution made in part I.A above. Add just enough NaF solid to cover the tip of a spatula into the test tube. Mix gently. Put a few mL of the liquid into two test tubes. Add a few mL of the KSCN solution to one of these test tubes and 3-6 drops of $Fe(NO_3)_3$ solution to the other. Note observations below.

What role does NaF play in what you observed? (i.e., how might it cause the color change?)

D. Fill two test tubes 1/3 full of the solution made in part I.A above. Place one of these test tubes in a hot water bath and the other in an ice bath. Leave the test tubes in the temperature baths until the temperature of the solutions becomes equalized with the baths (5-10 minutes). Remove the test tubes and compare the solutions with the standard you made in part I.B above. What effect does temperature have on the solution? What is the meaning of the color change.

II. $CuSO_4$ + NH_3

A. Put about 2 mL of $CuSO_4$ (copper (II) sulfate solution labeled 0.1 M) in a test tube. Note the appearance of the solution.

B. Using a medicine dropper add NH_3 (ammonia solution labeled 1 M) to the $CuSO_4$ dropwise mixing the solution between drops. Make observations and continue adding the NH_3 by the dropperful until the total volume is approximately 6 mL. Record your observations below.

The reactions you observed were: $Cu^{2+}(aq) + 2OH^-(aq) \longrightarrow Cu(OH)_2(s)$

$Cu^{2+}(aq) + 4NH_3(aq) \longrightarrow Cu(NH_3)_4^{2+}(aq)$

Identify the colors of each reactant and product of the above reactions.

C. Add HCl (hydrochloric acid labeled 1.5M) dropwise to the solution in part B. Make observations and continue adding the HCl until 3 mL have been added. What role does HCl play in the Cu^{2+} reaction? What relationship does it have with the NH_3 solution? Write an equation representing this relationship. (Hint: although a water solution of ammonia is given the chemical formula, NH_3, the solution actually contains a number of chemical materials including NH_3, NH_4^+, and OH^-.)

D. Add the NH_3 dropwise to the solution in part C. Make observations and continue adding the NH_3 until the total volume is approximately 15 mL. Record your observation below.

E. Generalize about the chemical reactions you studied in parts I and II. What did your observations tell you about reactions of these types?

IRON (III) NITRATE AND POTASSIUM THIOCYANATE
Exp. I-5

Name_____ Lab Section_____

Lab Partner _____

Pre-lab assignment: Review the textbook discussion of Molar (mole/liter) concentration units. Study Appendix D.1 on buret volume measurements and Appendix D.2 on the theory and use of spectrophotometers.

I. A Qualitative Look at a Reaction.

A. Dissolve about 1 cm³ of hydrated iron (III) nitrate, $Fe(NO_3)_3(H_2O)_9$, in 20 mL of distilled water. Dissolve about 1 cm³ of potassium thiocyanate, KSCN, in another 20 mL of water. Label each solution and describe the appearance of each.

B. Mix about 10 mL of each solution in a third beaker. Describe the results. (Set aside the remaining 10 mL of the $Fe(NO_3)_3$ and KSCN solutions for use in part III.C).

C. Assuming the reaction involves the combining of dissolved ions, what possible combinations of ions could be causing the reaction? Devise and conduct experiments that will distinguish between the possibilities. Describe the results of your experiments and state your conclusions concerning the ions actually reacting. Briefly explain your reasoning.

II. A Quantitative Look at the Reaction: $Fe^{3+}(aq) + SCN^-(aq) \longrightarrow FeSCN^{2+}(aq)$

You will be using a spectrophotometer to measure the color intensity and thus the concentration of the $FeSCN^{2+}(aq)$ formed in several solutions. Since the experiments depend on concentrations, it is important for the precision of the results that the volume measurements be made with care. Review Appendix D.1 on the proper use of burets. Also, you should be careful to not unwittingly dilute or contaminate your solutions.

A. Label eight dry 150 mm test tubes as numbers 1 to 8. Obtain in separate, dry, labeled beakers 80 mL of KSCN solution and 80 mL of $Fe(NO_3)_3$ solution. Record the exact molar concentration of each solution in the Table on page 157. (Each solution contains HNO_3 at 0.3 M to prevent the formation of iron hydroxides.)

Clean two burets, rinse one with two 5 mL portions of the KSCN, rinse the other with $Fe(NO_3)_3$ solution. Fill each buret with the appropriate solution making sure the tip is free of air gaps. Rinse and fill a third buret with distilled water. Label each buret.

B. You are to prepare eight solutions of the two reactants such that each contains the same amount of KSCN but differing amounts of $Fe(NO_3)_3$ in 20.00 total mL. This will be done by carefully buretting amounts of the stock solutions into the test tubes. The amounts suggested are given in the following table.

Test Tube	mL Stock KSCN	mL Stock $Fe(NO_3)_3$	mL H_2O
1	5.00	0.50	14.50
2	5.00	1.00	14.00
3	5.00	1.50	13.50
4	5.00	2.00	13.00
5	5.00	2.50	12.50
6	5.00	10.00	5.00
7	5.00	13.00	2.00
8	5.00	15.00	0.00

Make up your solutions recording the exact volumes used (to $\pm$ 0.01 mL) in the table on page 157. The order of addition of the reagents is not important.

Mix each solution by stoppering with a rubber stopper and inverting the tube several times. Shaking will not properly mix the reagents. Be sure to wipe the stopper dry between use with different test tubes.

C. Read the color intensity (Absorbance) of the $FeSCN^{2+}$ in each solution and record it in the table. In using the spectrophotometer follow the directions provided by the instructor and/or those given in Appendix D.2. The light wavelength of 447 nm is suggested for the measurements.

Retain at least 5 mL of test solution # 2 for use in part III.C. If a spectrophotometer is not immediately available for your use, do parts III.A, B and C while you are waiting.

DATA TABLE

Molarity of stock KSCN solution _____M

Molarity of stock $Fe(NO_3)_3$ solution _____M

Test Tube	mL Stock KSCN	mL Stock $Fe(NO_3)_3$	mL H_2O	Total added SCN^- conc (M)	Total added Fe^{3+} conc (M)	Absorbance at 447 nm
1	_____	_____	_____	_____	_____	_____
2	_____	_____	_____	_____	_____	_____
3	_____	_____	_____	_____	_____	_____
4	_____	_____	_____	_____	_____	_____
5	_____	_____	_____	_____	_____	_____
6	_____	_____	_____	_____	_____	_____
7	_____	_____	_____	_____	_____	_____
8	_____	_____	_____	_____	_____	_____

III. A. Calculate the total M concentration of SCN⁻ and Fe^{3+} added to each test solution. Enter these in the table on page 157. Show the calculations for test solution # 5 below.

B. Considering that fact that Fe^{3+} and SCN^- react 1:1 to form the colored ion $FeSCN^{2+}$, what is the **maximum** mole/liter concentration of $FeSCN^{2+}$ that could form in each test solution? Explain your reasoning below. (Hint: What is the limiting reagent in each solution?)

C. Into each of 3 small test tubes put 10 drops (1/2 mL) of test solution # 2. Into one put 5 drops of the KSCN solution you prepared in part I.A. Into the second test tube put 5 drops of the $Fe(NO_3)_3$ solution from part I.A; into the third put 5 drops of H_2O. Mix the solutions and compare them. Describe the results. What do the results indicate about the completeness of the reaction in test solution # 2? Briefly explain your reasoning.

D. What conclusions can be drawn from the data on page 157 about the completeness of the reaction in each test solution? Offer an explanation for your findings. It may be helpful to graph the absorbance vs. Fe^{3+} data. (Hints: Since Absorbance = constant x $[FeSCN^{2+}]$, how would the absorbances compare if the reaction went to completion in each solution? How can the actual absorbances be explained? Why might the reaction apparently go to completion in some situations but not in others?)

IV. A. To analyze the data mathematically, we need to determine the constant in the equation:

$$Abs = constant \times [FeSCN^{2+}]$$

This can be done by inspecting the graph in part III.D. If this graph has a plateau, we can assume these solutions are about 100% reacted. Thus their $[FeSCN^{2+}]$ can be calculated.

Using this approach, determine a value for the constant in the equation above. Then use this value with the non-plateau solutions and compute $[FeSCN^{2+}]$, $[SCN^-]$, and $[Fe^{3+}]$. Enter the data in the following table. Show one of these calculations in the space below.

Test Tube	Actual Concentrations (M) of			Results: Possible Combinations		
	$FeSCN^{2+}$	SCN^-	Fe^{3+}			
1	_____	_____	_____	_____	_____	_____
2	_____	_____	_____	_____	_____	_____
3	_____	_____	_____	_____	_____	_____
4	_____	_____	_____	_____	_____	_____
5	_____	_____	_____	_____	_____	_____
6	_____	_____	_____	_____	_____	_____
7	_____	_____	_____	_____	_____	_____
8	_____	_____	_____	_____	_____	_____

B. What conclusions can be drawn from these data? Find an algebraic equation that relates the actual concentrations of $FeSCN^{2+}$, SCN^- and Fe^{3+} to each other in the non-plateau solutions. (Hint: try all possible combinations of the three concentrations by multiplication and/or division. For example multiply all three together, multiply two and divide by the third, etc., looking for the combination that gives the most constant result.) Summarize your results in the table above. Be sure to label each column. Discuss your results below.

OXIDATION-REDUCTION REACTIONS
AND VOLTAIC CELLS—Exp. J-1

Name_____ Lab Section_____

Lab Partner _____

I. Interaction of metals and metal ions

   A. Clean and rinse with distilled water, four 12x75 mm test tubes (small test tubes). Shake excess water out of the tubes. Fill each of the tubes half full of $Zn(NO_3)_2$ solution. Clean strips of Zn, Cu, Pb, and Ag metals by gently sandpapering the last one inch of each strip. Submerge the cleaned end of the strips into the solutions for about 30 seconds. Record your observations in the following table.

   B. Rinse the four test tubes and clean the metal strips. Fill the tubes half full of $Cu(NO_3)_2$ solution and place the metal strips in the solutions. Record your observations.

   C. Repeat the above procedure with $AgNO_3$ solution and then with $Pb(NO_3)_2$ solution.

	Zn(s)	Cu(s)	Pb(s)	Ag(s)
Zn^{2+}(aq)				
Cu^{2+}(aq)				
Ag^{+}(aq)				
Pb^{2+}(aq)				

D. Write balanced chemical equations for each system tested. (If you observed no change in the system write NO REACTION.) Pick two of these reactions and explain why you chose the products you did.

E. What patterns are shown in these data? (Hint: compare the relative reactivity of the metals. Compare the relative reactivity of the metal ions. Identify any connections between these comparisons.)

II. Voltaic Cells

A. Fill a 100 mL beaker 1/3 full of 1 M NH_4NO_3. Obtain a U-tube with a length of cord inside it and invert this system into the solution.

B. Clean, rinse with distilled water, and shake dry four 12x75 mm test tubes. Label the test tubes "Zn," "Pb," "Cu," and "Ag." Fill the test tubes 1/2 full of $Zn(NO_3)_2$, $Pb(NO_3)_2$, $Cu(NO_3)_2$, and $AgNO_3$ solutions respectively. Put a cleaned strip of Zn metal in the $Zn(NO_3)_2$ solution, a cleaned strip of Pb in the $Pb(NO_3)_2$ solution, a cleaned Cu wire in the $Cu(NO_3)_2$ solution, and a cleaned strip of Ag in the $AgNO_3$ solution. Bend the end of the metal strips over the edges of the test tubes. Store the test tubes in a 100 mL beaker.

C. Assemble a Zn/Pb cell. Do this by placing the Zn test tube (called the Zn half cell and indicated by the symbol Zn/Zn^{2+} | |) and Pb test tube (called the Pb half cell and indicated by the symbol Pb/Pb^{2+} | |) into a 100 mL beaker. Remove the U-tube from the NH_4NO_3, blot the strings on a paper towel, and connect the two test tubes by immersing one leg of the U-tube in each of the two test tubes.

D. Set a volt meter to measure 0-2 volts. Clip one lead to the Zn metal strip and the other to the Pb strip. Switch the meter on or connect the meter just long enough to take a stable reading. Then turn the meter off (or disconnect it).

E. Assemble each of the possible six cells from the four half cells, and record the voltage of each cell. Rinse (by dipping the ends of the U-tube in water and then in NH_4NO_3) and blot the U-tube between readings.

F. Go through the entire procedure again obtaining a second reading for each combination of half cells. Record these readings and compute an average reading for each combination of half cells.

Cell = 1/2 cell (1) + 1/2 cell (2)	**Voltage Readings**		
	First	**Second**	**Average**
1. $Zn/Zn^{2+} \parallel Pb^{2+}/Pb$	_____	_____	_____
2. $Zn/Zn^{2+} \parallel Cu^{2+}/Cu$	_____	_____	_____
3. $Zn/Zn^{2+} \parallel Ag^{+}/Ag$	_____	_____	_____
4. $Ag/Ag^{+} \parallel Pb^{2+}/Pb$	_____	_____	_____
5. $Ag/Ag^{+} \parallel Cu^{2+}/Cu$	_____	_____	_____
6. $Cu/Cu^{2+} \parallel Pb^{2+}/Pb$	_____	_____	_____

G. Electricity is usually viewed as a flow of electrons. Explain why these experimental set-ups cause an electrical current to flow through the wires and volt meter. (Hint: choose one combination of cells and suggest what reactions may be occurring in each half cell. How do the reactions of the two half cells combine to give a net flow of electrons? What might be the net reaction of the combined half cells?)

H. What patterns exist in the voltage readings for the several combinations of half cells? How do these patterns compare to those found in part 1.E?

I. What relationships are there among the voltages of the $Zn/Zn^{2+} \parallel Pb^{2+}/Pb$, $Zn/Zn^{2+} \parallel Cu^{2+}/Cu$, and $Pb/Pb^{2+} \parallel Cu^{2+}/Cu$ cells? Explore the relationships of other similarly related cells.

III. A. The Cu^{2+}/Cu half cell will be used in this experiment. You will also need Ag^+ solutions of the following molarities: 0.20 M (solution from the previous experiment), 0.020 M, 0.0020 M and 0.00020 M. These last three can be prepared by a technique called serial dilution. The 0.020 M solution is prepared by diluting 1.0 mL of 0.20 M Ag^+ with 9.0 mL of distilled water. The 0.0020 M solution is prepared by diluting 1.0 mL of the 0.020 M Ag^+ with 9.0 mL of distilled water. In a similar manner the 0.00020 M Ag^+ solution is prepared from the 0.0020 M Ag^+ solution.

Place a few mLs of each Ag^+ solution in separate, clean, dried, labeled 12x75 mm test tubes.

B. Place the cleaned Ag strip in the 0.00020 M Ag^+ solution and connect this half cell to the 0.2 M Cu^{2+}/Cu half cell. Measure the voltage as described in part II. Record the data in the following table.

Remove the Ag metal strip and salt bridge, rinse them both and construct a cell using the 0.0020 M Ag^+ solution. Measure its voltage. Repeat this procedure with the 0.020 M and 0.20 M Ag^+ solutions.

C. Obtain a second reading of the voltage for each cell. Record these in the table and compute an average reading for each cell.

	Voltage Readings		
Cell	First	Second	Average
1. $Ag/0.00020$ M Ag^+ || Cu^{2+}/Cu	_____	_____	_____
2. $Ag/0.0020$ M Ag^+ || Cu^{2+}/Cu	_____	_____	_____
3. $Ag/0.020$ M Ag^+ || Cu^{2+}/Cu	_____	_____	_____
4. $Ag/0.20$ M Ag^+ || Cu^{2+}/Cu	_____	_____	_____

D. What conclusion can be drawn from these data? Can the pattern of the data be expressed as a simple algebraic equation? Justify your answers. It might be helpful to graph the data.

ELECTROLYSIS REACTIONS—Exp. J-2

Name_____ Lab Section_____

Lab Partner _____

I. Lead and Copper Reactions

A. Obtain approximately 50 mL of 0.2M solutions of $Cu(NO_3)_2$ and $Pb(NO_3)_2$ and place each in separate 100 mL beakers. Clean a copper and lead electrode by sanding lightly one inch of the bottom tip. Place the lead electrode in the $Cu(NO_3)_2$ solution and the copper electrode in the $Pb(NO_3)_2$ solution. Let them remain in solution for 5 minutes. Record any observations.

Write chemical equations for any changes you noticed.

B. Place a second Cu electrode in the $Pb(NO_3)_2$ solution separated from the first Cu electrode. Attach a D.C. current source to the electrodes. Note which electrode is attached to which pole of the source. Follow your instructor's instructions about how to use your D.C. current source. Your apparatus should look like the diagram. Switch on the power and allow it to run for approximately 5 minutes.

Disconnect and turn off the power supply. Record your observations.

Write chemical equations for any changes you noticed.

C. Compare your results in part I.B with those of part I.A. How would you account for any differences or similarities? What is the role of your power supply?

II. The Copper and H^+ Reaction—Qualitative

A. Fill a test tube full of distilled water and invert it into a 250 mL beaker with about 120 mL of distilled water such that no air is trapped in the test tube. Carefully add with stirring 40 mL of 4 M H_2SO_4. Connect a power supply to Cu electrodes in an apparatus similar to the one shown in part I.B (except H_2SO_4 solution has replaced $Pb(NO_3)_2$ solution). Be careful to note which electrodes are attached to the negative and positive poles. Run the reaction for about 5 minutes. Record your observations:

B. Collect some of the gas being produced by capturing bubbles in the filled inverted test tube. When filled with gas remove the test tube carefully holding it upside down. Hold a lighted match to its opening. Record your observations and speculate on the composition of the gas (see table C-2 in Appendix C if you're not sure how to do this).

C. Write a half-reaction describing the chemistry at the electrode connected to the negative pole of the power supply. (Use table C-6 in Appendix C for possible half-reactions. The half-reactions in table C-6 may be reversed if that will better describe the reaction being studied.)

Write a half-reaction describing the chemistry at the electrode connected to the positive pole of the power supply.

D. Combine the two half-reactions in part C to describe the overall reaction occurring in the beaker. Draw a diagram of the apparatus tracing the flow of electricity.

III. The copper and H^+ Reaction—Quantitative

A. Obtain and clean a 50 mL buret. Using a 10 mL graduate, measure the volume contained by the buret between the 50 mL mark and the stopcock. Record this value on the data table on page 171.

Put about 375 mL of distilled water in a 600 mL beaker. Carefully add 125 mL of 4 M H_2SO_4 and stir. The solution will be 1 M. (Your instructor may have an alternative method for obtaining this solution.)

Invert the buret into the solution and carefully fill it by opening the stopcock and drawing the solution into the buret using a suction bulb. Close the stopcock when the space below the stopcock is just filled. Leave the tip empty. Check the system for leaks.

Obtain a length of copper wire with the ends stripped and the middle covered with water-tight plastic. Bend and place the wire so that one stripped end is entirely inside the buret and the other end is outside the beaker (see diagram).

169

B. Clean a Cu electrode by sanding, dipping it into 6M nitric acid, rinsing it with water, and drying it. Weigh the electrode as accurately as is possible with the balances available to you. Record the value on the data table.

Bend the electrode and place it in the solution. Connect the leads from the Cu electrode to the + pole of the power source/ammeter and the − pole to the copper wire. (see diagram) Note the time to the nearest second and quickly adjust the current to between 0.10 amp and 0.15 amp according to the ammeter. Record the time and current on the data table. Check the current every 10 minutes and adjust if necessary.

Obtain an atmospheric pressure and temperature reading and record these on the data table.

C. Six times during the progress of the reaction, record on the data table the volume reading and the elapsed time. Be careful when reading the graduations on the buret to note they are upside down. Determine the actual volume of gas collected and record this on the table.

D. Sometime before the collected gas reaches the zero mark on the buret turn off or disconnect the power noting the exact time to the second. Also note the elapsed time and the final volume reading. Record these data on the data table.

E. Very carefully remove the Cu electrode from the solution. Dip the electrode into a beaker of distilled water being careful not to dislodge any material. Dip the electrode into acetone and then allow it to air dry until no more moisture remains. Weigh the electrode and record this data on the data table.

Data Table

Volume of buret
    between 50 mL ⟶ stopcock _____mL

Initial mass Cu _____g   Final mass Cu _____g

Time began _____   Time finished _____

Current _____amps

Atmospheric Pressure _____torr

Temperature _____°C

Volume Reading	Time Elapsed (sec)	Volume Gas Collected (mL)
_____	0	0
_____	_____	_____
_____	_____	_____
_____	_____	_____
_____	_____	_____
_____	_____	_____
_____	_____	_____
final reading _____	_____	_____

IV. Analysis

  A. What is the relationship between the volume of gas and the elapsed time? Develop an algebraic equation which expresses this relationship. It might be helpful to graph your data and develop the algebraic equation from the graph.

B. Why are volume of gas and time related in the manner shown in part IV.A? (Hint: how is the current related to production of gas according to the 1/2 reactions developed in part II? How is this related to your equation in section IV.A?)

C. From your P, V, T data, calculate the number of moles of gas produced during the experiment. Show your work. (Don't forget to adjust the partial pressure of your gas sample to allow for the presence of water vapor.) The vapor pressure of water at various temperatures can be found in the *Handbook of Chemistry and Physics*.

D. From the weight data, calculate the number of moles of Cu consumed during the experiment. Show your work.

E. Compare the amount of H_2 and Cu produced or consumed during the experiment. How does this relate to the amount of electrons utilized? How does this relate to your proposed chemical equation from part II.D?

F. Write a generalization which summarizes your findings in this section of the experiment. (Hint: how much silver would be produced (or consumed) if it was used instead of Cu in the cell? See the 1/2 reactions in table C-6 in Appendix C)

G. (Optional) The quantity of electricity (coulombs) = current (amps) x time (sec). Calculate the relationship between the quantity of electricity and amount of electrons in an electrochemical reaction.

BLEACH AND BROMOCRESOL GREEN—Exp. K-1

Name_____ Lab Section_____

Lab Partner _____

Precautions

NaOCl(aq) is liquid bleach. Its vapors are irritating and prolonged contact can be irritating to skin. Avoid breathing or touching the undiluted stock solution. Bromocresol green is a dye and can permanently stain your clothes.

I. Concentration Effects

    A. Obtain about 100 mL of pH 7 buffer in a 250 mL beaker, about 20 mL of pH 4 buffer in a 50 mL beaker, and 50 mL of bromocresol green (BCG) solution.

    B. Into a 100 mL beaker measure 20 mL of a pH 4 buffer, 30 mL of distilled water, and 2 mL of BCG solution. Swirl the beaker to mix. Your solution should be yellow with a hint of green in it. If not see your laboratory instructor for help. This solution is to be used as a standard for the following experiments.

    C. Measure 7 mL of 5.25% NaOCl (liquid bleach) with a 10 mL graduated cylinder. Combine this with 413 mL of distilled water (measured with a 50 mL graduated cylinder) in a 600 mL beaker. Mix well.

    D. Into a 100 mL beaker measure 10 mL of a pH 7 buffer, 10 mL of bleach solution (made in part I.C), and 30 mL of distilled water. Into a 10 mL graduated cylinder measure exactly 3 mL of BCG solution. Arrange the 100 mL beaker, and the standard solution next to each other on a piece of white paper. With a stopwatch begin timing at the same time that the BCG is added to the 100 mL beaker. Swirl the solution to mix and observe the color of the solution. Stop timing as soon as the colors of the solution and the standard match. The match should be one of *color*, not of brightness or intensity. Repeat this procedure. If the amount of time for the color change is constant (or almost constant), there is no need to repeat a third time, otherwise repeat until two readings are consistant. Record your data in the table.

E. Using the procedure in section I.D, measure the amount of time for solutions to match for each of the following combinations and any other combinations you want to try. The combination of volumes of bleach, buffer, and water should add up to 50 mL.

Exp. #	Bleach soln(mL)	pH 7 Buffer(mL)	Water(mL)	BCG(mL)	Time(sec)	Time(sec)	Time(sec)
1	10	10	30	3			
2	10	10	30	6			
3	20	10	20	3			
4	30	10	10	3			
5	40	10	0	3			
6	20	5	25	3			
7							
8							

F. What role did the buffer play in the rate of the reaction? How did you arrive at this conclusion?

G. How did the concentration of bromocresol green effect the rate of the reaction? What experiments did you use to come to your conclusion (use the numbers from the above table)?

H. How is the rate of the reaction affected by the concentration of the bleach solution? It might be helpful to graph your data. What is the effect of doubling the concentration?

What experiment numbers would be useful in a graph demonstrating the effect of concentration of bleach on the rate of the reaction? If you eliminated any experiments explain why.

II. Temperature Effects

A. Using the same procedure as part I above, make a solution of 10 mL of bleach solution (made in part I.C), 10 mL of pH=7 buffer, and 30 mL of water in a 100 mL beaker.

B. Place a 250 mL beaker 1/4 filled with water on a ring stand over a piece of gauze and heat the water with a bunsen burner. Place the 100 mL beaker with the test solution in the hot water bath. With a thermometer measure the temperature of the test solution. When the temperature is approximately 10°C higher than room temperature, remove the beaker from the water bath with tongs.

C. Place the beaker next to the standard solution and add 3 mL of BCG while beginning timing. When the reaction has finished measure the temperature of the solution. (Record the temperature at the color change.)

D. Using the procedure outlined above, measure the amount of time for the solution to match the standard for six different temperatures ($15°C\pm$ room temp). Temperatures below room temperature can be obtained by using an ice bath.

	Time(sec)	Initial Temperature	Final Temperature	Average Temperature
1				
2				
3				
4				
5				
6				

E. How did the temperature of the solution affect the rate of the reaction? It might be helpful to graph your data. What temperature change resulted in doubling the rate?

THE BROMINATION OF ACETONE—Exp. K-2

Name_____ Lab Section_____

Lab Partner _____

Pre-lab assignment: Study Appendix D.2 on the use and theory of spectrophotometers for measuring color intensity. You will read %T as it is easier to see. You will then convert it mathematically to absorbance.

I. $(CH_3)_2CO(aq) + Br_2(aq) \longrightarrow BrCH_2COCH_3(aq) + HBr(aq)$

A. Obtain 50 mL of 0.02M Br_2 solution. Store it in a clean labeled container.

B. Make 100 mL of a 4.0 M acetone solution by measuring 30.0 mL of reagent acetone using appropriate graduated cylinders or pipets. Place the acetone in a clean 100 mL volumetric flask. Fill the flask half full of distilled water and mix. Finish diluting to the 100 mL mark and mix. Transfer the solution to a clean labeled container.

C. Make 100 mL of a 1.0 M H^+ solution by measuring 25 mL of 4.0 M HCl. Half fill a clean, rinsed 100 mL volumetric flask with distilled water. Add the measured HCl. Finish diluting to the 100 mL mark and mix. Transfer the solution to a clean labeled container and allow the solution to come to room temperature.

D. Measure 5 mL of each of the three solutions into separate 10 mL graduate cylinders. Also measure 10 mL of distilled water. Mix the four samples together and observe for approximately five minutes. Record your observations.

E. Speculate on what caused the change noticed.

II. Data Collection

Note: The yellow color of the solution absorbs its maximum light at a wave length of 395 nm. Consequently this is the wave length which will be used on the Spectrophotometer.

A. Warm up the spectrophotometer, set the wave length at 395 nm, zero the transmission setting, and using a tube of distilled water set the 100% transmission.

B. You will do several experiments varying the concentrations of Br_2, acetone, and H^+ according to the table in section C. For each experiment, use the following procedure:

(1) Measure the proper amount of each solution and water (total volume in each case = 25 mL).

(2) Swirl mix all solutions except Br_2 in a rinsed, clean, and towel dried 100 mL beaker.

(3) Add the Br_2 solution. Begin timing as solution is added.

(4) Swirl mix.

(5) Rinse a spectrophotometer tube with the solution by 1/2 filling the tube and dumping the solution in a waste container (600 mL beaker). Do this twice.

(6) Fill (at least 1/2 full) the tube and wipe the tube clean with a tissue holding the tube at the top.

(7) Place the tube in the spectrophotometer and close the light cover.

(8) Steps 1-7 should be done in less than 45 sec.

(9) Take % transmission reading at 45 sec., 60 sec., 90 sec., 120 sec., 150 sec., and 180 sec. or until the reaction is finished if sooner than 180 sec.

(10) Dump the spectrophotometer tube. Dump the 100 mL beaker. Rinse the beaker three times with distilled water and dry with a tissue.

(11) Begin with the next sample.

(12) Occasionally check the 0 and 100% transmission settings.

This process is much more easily accomplished if all your materials are laid out in an organized fashion. Divide duties for partners.

C. Experiments:

Experiment #	mL .02M Br_2	mL 4 M Acetone	mL 1M H^+	mL Distilled Water
1	5	5	5	10
2	10	5	5	5
3	10	5	10	0
4	10	10	5	0

D. Record the data for the four experiments in the following tables. Calculate the absorbance. (A = −logT, T is the decimal fraction of %T).

Experiment 1

%T	time (sec)	A
_____	45	_____
_____	60	_____
_____	90	_____
_____	120	_____
_____	150	_____
_____	180	_____

Experiment 2

%T	time (sec)	A
_____	45	_____
_____	60	_____
_____	90	_____
_____	120	_____
_____	150	_____
_____	180	_____

Experiment 3

%T	time (sec)	A
_____	45	_____
_____	60	_____
_____	90	_____
_____	120	_____
_____	150	_____

Experiment 4

%T	time (sec)	A
_____	45	_____
_____	60	_____
_____	90	_____
_____	120	_____
_____	150	_____

III. Data Analysis

A. The rate of this reaction can be expressed as $-\dfrac{\Delta[Br_2]}{\Delta t}$. Since $A \propto [Br_2]$ the rate can be expressed as $-\dfrac{\Delta A}{\Delta t}$. If a plot of A (y-axis) vs. t (x-axis) is a straight line, then the ratio $-\dfrac{\Delta A}{\Delta t}$ is the slope of this line. Calculate the rate of each experiment. Show your work for experiment #1. Record your data in the table in part B.

B. Complete the following table for the initial concentrations of the 25 mL solution.

Experiment	$[Br_2]$	[Acetone]	$[H^+]$	rate $= -\dfrac{\Delta A}{\Delta t}$
1	_____	_____	_____	_____
2	_____	_____	_____	_____
3	_____	_____	_____	_____
4	_____	_____	_____	_____

C. What can be said about the effect each of the three solutions has on the rate of the reaction? (Hint: use combinations of experiments to prove your point and isolate the effect of each solution.)

D. An equation called a rate law shows the relationship between the concentration of reactants and the rate of the reaction. The rate law for the reaction studied in this experiment has the form:

$$\text{rate} = \frac{-\Delta A}{\Delta t} \propto -\frac{[\Delta Br_2]}{\Delta t} = k[Br_2]^x[Acetone]^y[H+]^z$$

By using the data from the proper experiments, concentration dependences can be isolated from the ratio:

$$\frac{\text{rate exp \# p}}{\text{rate exp \# q}} = \frac{k[Br_2]_p^x[Acetone]_p^y[H+]_p^z}{k[Br_2]_q^x[Acetone]_q^y[H+]_q^z}$$

where p and q represent two of the four experiments (chosen so that all of the terms representing x, y, and z can be cancelled but one). Determine values for x, y, and z. Show your work.

E. Use the rate law determined in part D to calculate a value for k for each of the four experiments. Show your work. Is k a constant?

Exp.#	k	Exp.#	k
1	_____	3	_____
2	_____	4	_____

QUALITATIVE ANALYSIS FOR Ag^+, Pb^{2+}, Ni^{2+}, Ba^{2+}, Fe^{3+}— L-1

Name_____ Lab Section_____

Lab Partner _____

Pre-lab assignment: Study Appendix D.4 on qualitative analysis techniques.

PART ONE

I. Obtain approximately 10 mL of 0.1 M solutions of Ag^+, Pb^{2+}, Ni^{2+}, Ba^{2+} and Fe^{3+} in separate clean, labeled beakers. (Each of these solutions were prepared from the nitrate salt of the cation.) You are to subject each of these solutions to the six tests outlined below and record your observations in the table provided.

II. Chemical Tests

1. Put 10 drops of test solution into a 75 mm test tube and then add 5 drops of 3 M H_2SO_4.

2. To 10 drops of fresh test solution add 5 drops of 6 M HCl. If a precipitate forms, mix the sample and centrifuge. Wash the precipitate twice with H_2O and save it for test 3.

3. Add 1 1/2 mL of fresh distilled water to the precipitate from test 2, mix. Heat the tube in a boiling water bath for 4 minutes. Mix frequently. Centrifuge and decant the hot liquid. To the hot liquid add 1 drop of K_2CrO_4.

4. To 10 drops of fresh test solution add 10 drops of 6 M NaOH. If no precipitate develops, tests 5 and 6 need not be done. If a precipitate forms, mix the sample and centrifuge. Discard the supernatant and save the precipitate for test 5.

5. Wash the precipitate from test 4 once with H_2O. Add 10 drops of 15 M NH_3 to the precipitate and mix for one minute. Centrifuge.

6. Repeat test 4 to obtain a hydroxide precipitate free of supernatant. Wash the precipitate once with H_2O. Add 5 drops of 6 M HCl to the precipitate and mix. Add 2 drops of 1 M NH_4SCN solution.

III. Observations

CATION

TEST and REAGENT	Ag^+	Pb^{2+}	Ni^{2+}	Ba^{2+}	Fe^{3+}
1. SO_4^{2-}					
2. Cl^-					
3. a. heat					
b. CrO_4^{2-}					
4. OH^-					
5. NH_3					
6. a. H^+					
b. SCN^-					

IV. Possible Equations

From your data complete the equations listed below. You will be speculating, but your speculation will be accurate in 90% of the cases if you use the following guidelines. (1) Assume that most reactions are a simple combination of cations and anions. (2) If the product is dissolved it is ionic. (3) If the product is solid it is neutral. If you observed no change when the particular reagents were mixed, write N.R. for "no reaction".

1. $Ag^+ + SO_4^{2-}$ ⟶ _____

2. $Pb^{2+} + SO_4^{2-}$ ⟶ _____

3. $Ni^{2+} + SO_4^{2-}$ ⟶ _____

4. $Ba^{2+} + SO_4^{2-}$ ⟶ _____

5. $Fe^{3+} + SO_4^{2-}$ ⟶ _____

6. $Ag^+ + Cl^-$ ⟶ _____ $\xrightarrow{+\ heat}$ _____ $\xrightarrow{+CrO_4^{2-}}$ _____

7. $Pb^{2+} + Cl^-$ ⟶ _____ $\xrightarrow{+\ heat}$ _____ $\xrightarrow{+CrO_4^{2-}}$ _____

8. $Ni^{2+} + Cl^-$ ⟶ _____ $\xrightarrow{+\ heat}$ _____ $\xrightarrow{+CrO_4^{2-}}$ _____

9. $Ba^{2+} + Cl^-$ ⟶ _____ $\xrightarrow{+\ heat}$ _____ $\xrightarrow{+CrO_4^{2-}}$ _____

10. $Fe^{3+} + Cl^-$ ⟶ _____ $\xrightarrow{+\ heat}$ _____ $\xrightarrow{+CrO_4^{2-}}$ _____

11. $Ag^+ + OH^-$ ⟶ _____ $\xrightarrow{+NH_3}$ _____

$Ag^+ + OH^-$ ⟶ _____ $\xrightarrow{+H^+}$ _____ $\xrightarrow{+SCN^-}$ _____

12. $Pb^{2+} + OH^-$ ⟶ _____ $\xrightarrow{+NH_3}$ _____

$Pb^{2+} + OH^-$ ⟶ _____ $\xrightarrow{+H^+}$ _____ $\xrightarrow{+SCN^-}$ _____

13. $Ni^{2+} + OH^-$ ⟶ _____ $\xrightarrow{+NH_3}$ _____

$Ni^{2+} + OH^-$ ⟶ _____ $\xrightarrow{+H^+}$ _____ $\xrightarrow{+SCN^-}$ _____

14. $Ba^{2+} + OH^-$ ⟶ _____ $\xrightarrow{+NH_3}$ _____

$Ba^{2+} + OH^-$ ⟶ _____ $\xrightarrow{+H^+}$ _____ $\xrightarrow{+SCN^-}$ _____

15. $Fe^{3+} + OH^-$ ⟶ _____ $\xrightarrow{+NH_3}$ _____

$Fe^{3+} + OH^-$ ⟶ _____ $\xrightarrow{+H^+}$ _____ $\xrightarrow{+SCN^-}$ _____

V. A. Assume you were given a solution that was said to contain Ba^{2+} cations only. Describe how you would confirm or deny the presence of this caption in the solution. Explain your reasoning.

B. Assume you were given a solution that was said to contain Pb^{2+} and Ag^+ cations only. Describe how you would determine if neither, only one or both of these ions were present in solution. Explain your reasoning.

C. Assume you were given a solution that was said to contain Fe^{3+} and Ni^{2+} cations. Describe how you would determine if neither, only one or both of these ions were present in solution. Explain your reasoning.

D. Assume you were given a solution that was said to contain Ag^+, Pb^{2+}, Ni^{2+}, Fe^{3+}, and Ba^{2+} cations only. Describe how you would confirm or deny the presence of each of these ions in the solution. Explain your reasoning. (Hint: You may wish to design an analysis scheme—a flow chart—such as the one shown in the Appendix on page 288.)

PART TWO

ANALYSIS OF AN UNKNOWN

You will be given 10 mL of a solution of unknown composition which might contain any or all of the cations: Ag^+, Pb^{2+}, Fe^{3+}, Ba^{2+}, and Ni^{2+}. You are to analyze your solution in whatever manner you feel is appropriate to determine unequivocally which of the cations are present.

The lab report for this section should include the following:

1. An experimental section. This should be a step-by-step accounting of the experiments conducted and the results observed. Include balanced ionic equations where appropriate.

2. A conclusion section. This should be a paragraph telling which cations you think were in your solution and telling why you reached this conclusion.

QUALITATIVE ANALYSIS FOR Zn^{2+}, Ca^{2+}, Cu^{2+}, Al^{3+}, Co^{2+}—Exp. L-2

Name_____ Lab Section_____

Lab Partner _____

Pre-lab assignment: Study Appendix D.4 on qualitative analysis techniques.

PART ONE

I. Obtain approximately 10 mL of 0.1 M solutions of Zn^{2+}, Ca^{2+}, Cu^{2+}, Al^{3+} and Co^{2+} in separate clean, labeled beakers. (Each of these solutions were prepared from the nitrate salt of the cation.) You are to subject each of these solutions to the five tests outlined below and record your observations in the table provided.

II. Chemical Tests

1. Put 10 drops of test solution into a 75 mm test tube. Test the pH. If needed, add 3 M NH_3 (NH_4OH) or 3 M HNO_3 dropwise until the pH is neutral; i.e., between 6 and 8. Now add 7 drops of 0.3 M Na_3PO_4.

2. To 10 drops of fresh test solution add 3 M HNO_3 until the solution is strongly acidic; i.e., pH < 3. Add 7 drops of 0.2 M $K_3Fe(CN)_6$, potassium ferricyanide.

3. To 10 drops of fresh test solution add 6 M NaOH until the solution is just basic; i.e., pH >8. Add 10 drops of 6 M NaOH. If a precipitate forms, mix the sample and centrifuge. Decant the supernatant and save the precipitate for test 4.

4. Wash the precipitate from test 3 twice with distilled water. Add 10 drops of 15 M NH_3 and mix for 1 minute. Centrifuge.

5. Repeat step 3 to obtain a hydroxide precipitate free of supernatant. Wash the precipitate twice with distilled water. Add 15 drops of 3 M HNO_3 to the precipitate and mix.

III. Observations

TEST and REAGENT	CATION				
	Zn^{2+}	Ca^{2+}	Cu^{2+}	Al^{3+}	Co^{2+}
1. PO_4^{3-}					
2. $Fe(CN)_6^{3-}$					
3. OH^-					
4. NH_3					
5. H^+					

IV. Possible Equations

From your data complete the equations listed below. You will be speculating, but your speculation will be accurate in 90% of the cases if you use the following guidelines. (1) Assume that most reactions are a simple combination of cations and anions. (2) If the product is dissolved it is ionic. (3) If the product is solid it is neutral. If you observed no change when the particular reagents were mixed, write N.R. for "no reaction".

1. $Zn^{2+} + PO_4^{3-}$ ⟶ _____

2. $Ca^{2+} + PO_4^{3-}$ ⟶ _____

3. $Cu^{2+} + PO_4^{3-}$ ⟶ _____

4. $Al^{3+} + PO_4^{3-}$ ⟶ _____

5. $Co^{2+} + PO_4^{3-}$ ⟶ _____

6. $Zn^{2+} + Fe(CN)_6^{3-}$ ⟶ _____

7. $Ca^{2+} + Fe(CN)_6^{3-}$ ⟶ _____

8. $Cu^{2+} + Fe(CN)_6^{3-}$ ⟶ _____

9. $Al^{3+} + Fe(CN)_6^{3-}$ ⟶ _____

10. $Co^{2+} + Fe(CN)_6^{3-}$ ⟶ _____

11. $Zn^{2+} + OH^-$ ⟶ _____ $\xrightarrow{+NH_3}$ _____

$Zn^{2+} + OH^-$ ⟶ _____ $\xrightarrow{H^+}$ _____

12. $Ca^{2+} + OH^-$ ⟶ _____ $\xrightarrow{+NH_3}$ _____

$Ca^{2+} + OH^-$ ⟶ _____ $\xrightarrow{+H^+}$ _____

13. $Cu^{2+} + OH^-$ ⟶ _____ $\xrightarrow{+NH_3}$ _____

$Cu^{2+} + OH^-$ ⟶ _____ $\xrightarrow{+H^+}$ _____

14. $Al^{3+} + OH^-$ ⟶ _____ $\xrightarrow{+NH_3}$ _____

$Al^{3+} + OH^-$ ⟶ _____ $\xrightarrow{+H^+}$ _____

15. $Co^{2+} + OH^-$ ⟶ _____ $\xrightarrow{+NH_3}$ _____

$Co^{2+} + OH^-$ ⟶ _____ $\xrightarrow{+H^+}$ _____

V. A. Assume you were given a solution that was said to contain Ca^{2+} cations only. Describe how you would confirm or deny the presence of this cation in the solution. Explain your reasoning.

B. Assume you were given a solution that was said to contain Zn^{2+} and Al^{3+} cations only. Describe how you would determine if neither, only one or both of these ions were present in solution. Explain your reasoning.

C. Assume you were given a solution that was said to contain Cu^{2+} and Co^{2+} cations only. Describe how you would determine if neither, only one or both of these ions were present in solution. Explain your reasoning.

D. Assume you were given a solution that was said to contain Zn^{2+}, Ca^{2+}, Cu^{2+}, Al^{3+}, and Co^{2+} cations only. Describe how you would confirm or deny the presence of each of these ions in the solution. Explain your reasoning. (Hint: you may wish to design an analysis scheme—a flow chart—such as the one shown in the Appendix on page 288.)

PART TWO

ANALYSIS OF AN UNKNOWN

You will be given 10 mL of a solution of unknown composition which might contain any or all of the cations: Zn^{2+}, Ca^{2+}, Cu^{2+}, Al^{3+}, and Co^{2+}. You are to analyze your solution in whatever manner you feel is appropriate to determine unequivocally which of the cations are present.

The lab report for this section should include the following:

1. An experimental section. This should be a step-by-step accounting of the experiments conducted and the results observed. Include balanced ionic equations where appropriate.

2. A conclusion section. This should be a paragraph telling which cations you think were in your solution and telling why you reached this conclusion.

QUALITATIVE ANALYSIS FOR Cl^-, I^-, PO_4^{3-}, SO_4^{2-}, and $C_2O_4^{2-}$ —Exp. L-3

Name_____ Lab Section_____

Lab Partner _____

Pre-lab assignment: Study Appendix D.4 on qualitative analysis techniques.

PART ONE

I. Obtain approximately 10 mL of 0.1 M solutions of Cl^-, I^-, PO_4^{3-}, SO_4^{2-} and $C_2O_4^{2-}$ in separate clean, labeled beakers. (Each of these solutions were prepared from the K^+ or Na^+ salt of the anion.) You are to subject each of these solutions to the six tests outlined below and record your observations in the table provided.

II. Chemical Tests

1. Put 10 drops of test solution into a 75 mm test tube. Add 6 M HNO_3 until the solution is acidic, i.e. pH < 6. Add 8 drops of 0.2 M $AgNO_3$. If a ppt forms, mix the sample and centrifuge. Decant the spn. and save the ppt. for test 2.

2. Wash the ppt from test 1 twice with distilled water. Add 10 drops of H_2O, 10 drops of 3 M NH_3 and mix for 2 minutes. Centrifuge if needed. Record your result. Separate out the spn. and add 6 M HNO_3 until the solution is acidic.

3. To 10 drops of fresh test solution add 3 M NH_3 dropwise until the solution is just basic; i.e., pH = 8-9. If the solution by itself is basic, add 1 drop of 3 M NH_3. Add 7 drops of 0.3 M $Ba(NO_3)_2$. If a precipitate forms, mix the sample and centrifuge. Decant the supernatant and save the ppt. test 4.

4. Wash the ppt. from test 3 twice with distilled water. Add 20 drops of 6 M HNO_3 and mix. Place in a gently boiling water bath for 4 to 7 minutes, mixing the sample about every 30 seconds.

5. To 10 drops of fresh test solution, add 4 drops of 6 M HNO_3 then add 5 drops of 0.15 M ammonium molybdate. Mix and let stand in a hot water bath for about 2 minutes.

6. To 10 drops of fresh test solution, add 4 drops of 6 MHNO_3 and just 1 drop of 0.01 M $KMnO_4$. Mix and let stand for 5 minutes. (The disappearance of the purple color indicates that MnO_4^- was reduced to Mn^{2+}.)

III. Observations

TEST and REAGENT	ANION				
	Cl^-	I^-	PO_4^{3-}	SO_4^{2-}	$C_2O_4^{2-}$
1. Ag^+					
2. a. NH_3 b. H^+					
3. Ba^{2+}					
4. H^+					
5. $NH_4(MoO_3)_4^+$					
6. MnO_4^-					

IV. Possible Equations

From your data complete the equations listed below. You will be speculating, but your speculation will be accurate in 90% of the cases if you use the following guidelines. (1) Assume that most reactions are a simple combination of cations and anions. (2) If the product is dissolved it is ionic. (3) If the product is solid it is neutral. If you observed no change when the particular reagents were mixed, write N.R. for "no reaction."

1. $Cl^- + Ag^+$ $\longrightarrow$ _____ $\xrightarrow{NH_3}$ _____ $\xrightarrow{H^+}$ _____

2. $I^- + Ag^+$ $\longrightarrow$ _____ $\xrightarrow{NH_3}$ _____ $\xrightarrow{H^+}$ _____

3. $PO_4^{3-} + Ag^+$ $\longrightarrow$ _____ $\xrightarrow{NH_3}$ _____ $\xrightarrow{H^+}$ _____

4. $SO_4^{2-} + Ag^+$ $\longrightarrow$ _____ $\xrightarrow{NH_3}$ _____ $\xrightarrow{H^+}$ _____

5. $C_2O_4^{2-} + Ag^+$ $\longrightarrow$ _____ $\xrightarrow{NH_3}$ _____ $\xrightarrow{H^+}$ _____

6. $Cl^- + Ba^{2+}$ $\longrightarrow$ _____ $\xrightarrow{H^+}$ _____

7. $I^- + Ba^{2+}$ $\longrightarrow$ _____ $\xrightarrow{H^+}$ _____

8. $PO_4^{3-} + Ba^{2+}$ $\longrightarrow$ _____ $\xrightarrow{H^+}$ _____

9. $SO_4^{2-} + Ba^{2+}$ $\longrightarrow$ _____ $\xrightarrow{H^+}$ _____

10. $C_2O_4^{2-} + Ba^{2+}$ $\longrightarrow$ _____ $\xrightarrow{H^+}$ _____

11. $Cl^- + NH_4(MoO_3)_4^+$ $\longrightarrow$ _____

12. $I^- + NH_4(MoO_3)_4^+$ $\longrightarrow$ _____

13. $PO_4^{3-} + NH_4(MoO_3)_4^+$ $\longrightarrow$ _____

14. $SO_4^{2-} + NH_4(MoO_3)_4^+$ $\longrightarrow$ _____

15. $C_2O_4^{2-} + NH_4(MoO_3)_4^+$ $\longrightarrow$ _____

16. $Cl^- + MnO_4^-$ $\longrightarrow$ _____

17. $I^- + MnO_4^-$ $\longrightarrow$ _____

18. $PO_4^{3-} + MnO_4^-$ $\longrightarrow$ _____

19. $SO_4^{2-} + MnO_4^-$ $\longrightarrow$ _____

20. $C_2O_4^{2-} + MnO_4^-$ $\longrightarrow$ _____

V. A. Assume you were given a solution that was said to contain PO_4^{3-} anions only. Describe how you would confirm or deny the presence of this ion in the solution. Explain your reasoning.

B. Assume you were given a solution that was said to contain SO_4^{2-} and $C_2O_4^{2-}$ anions only. Describe how you would determine if neither, only one or both of these anions were present in solution. Explain your reasoning.

C. Assume you were given a solution that was said to contain Cl^- and I^- anions only. Describe how you would determine if neither, only one or both of these anions were present. Explain your reasoning.

D. Assume you were given a solution that was said to contain Cl^-, I^-, PO_4^{3-}, SO_4^{2-}, and $C_2O_4^{2-}$ anions. Describe how you would confirm or deny the presence of each of these ions in the solution. Explain your reasoning. (Hint: you may wish to design an analysis scheme—a flow chart—such as the one shown in the Appendix on page 288.)

PART TWO

ANALYSIS OF AN UNKNOWN

You will be given 10 mL of a solution of unknown composition which might contain any or all of the anions: Cl^-, I^-, PO_4^{3-}, $C_2O_4^{2-}$ and SO_4^{2-}. You are to analyze your solution in whatever manner you feel is appropriate to determine unequivocally which of the anions are present.

The lab report for this section should include the following:

1. An experimental section. This should be a step-by-step accounting of the experiments conducted and the results observed. Include balanced ionic equations where appropriate.

2. A conclusion section. This should be a paragraph telling which anions you think were in your solution and telling why you reached this conclusion.

IRON ANALYSIS—Exp. M-1

Name_____ Lab Section_____

Lab Partner _____

Pre-lab assignment: Study Appendix D.1 on titration techniques.

PART ONE

I. You will be investigating the quantitative analysis of iron ions by titration with permanganate, MnO_4^-. However, in these first parts, you will be performing qualitative tests to determine the chemical equations involved in the analysis. It is best to perform these tests over a white background (paper or towel).

Obtain about 20 mL of 0.1 M $FeSO_4$ (Fe^{2+} or ferrous ions) and 20 mL of 0.1 M $Fe(NO_3)_3$ (Fe^{3+} or ferric ions) in distilled water rinsed, labeled beakers. Also obtain about 20 mL of 2 M H_2SO_4, 5 mL of 0.03 M $K_2Cr_2O_7$, 5 mL of 1 M NH_4SCN, and 5 mL of 0.02 M $KMnO_4$ in clean labeled beakers.

A. Place about 5 mL of water, 5 mL of Fe^{2+} solution and 5 mL of Fe^{3+} solution in separate rinsed, labeled large test tubes or small beakers. Record the appearance of each in the table below. Add 1 mL of 2 M H_2SO_4 (about one dropperful) to each and record their appearance. Finally, add 4 drops of 0.03 M $K_2Cr_2O_7$ to each, mix, and record their appearance.

B. Using 1 mL of H_2SO_4, acidify fresh 5 mL samples of H_2O, Fe^{2+}, and Fe^{3+} solution. Add 1 drop of 1 M NH_4SCN to each solution and record your observations.

	Water blank	Fe^{2+} ion	Fe^{3+} ion
No Addition			
acid			
$Cr_2O_7^{2-}$			
SCN^-			

C. Explain briefly how $Cr_2O_7^{2-}$ and SCN^- can be used to test any solution for the presence of Fe^{2+} and/or Fe^{3+}?

II. The Reaction of Iron Ions with MnO_4^-.

A. [Note: when MnO_4^- decolors, it is being converted to Mn^{2+}.]

Acidify fresh 5 mL samples of H_2O, Fe^{2+}, and Fe^{3+} with 1 mL portions of H_2SO_4. Add 15 to 20 drops of 0.02 M MnO_4^- to each and mix. Record your observations. Save the solutions.

Water blank:

Fe^{2+} solution:

Fe^{3+} solution:

B. If any of these solutions are colorless, we can test them for the type of iron ions they now contain. Split any colorless iron ion solution into two parts. Add 4 drops of $Cr_2O_7^{2-}$ to one part and 1 drop of SCN^- to the other. Record your observations.

C. Do Fe^{2+} ions react with MnO_4^-? If you feel a reaction occurs, identify the products of the reaction. Explain in detail how your data support your conclusions.

Write a balanced chemical equation for the reaction of Fe^{2+} with MnO_4^-.

D. Do Fe^{3+} ions react with MnO_4^-? If you feel a reaction occurs, identify the products. Explain in detail how your data support your conclusions.

Write a balanced chemical equation for the reaction of Fe^{3+} with MnO_4^-.

III. The Titration Indicator

Permanganate, MnO_4^-, has such an intense color that it is often used as its own indicator in a titration. Consider again the colors of the solutions in part II.A and the reactions you deduced in parts II.C and II.D. Explain, in detail, how MnO_4^-'s color can be used to indicate the end point in a titration of Fe^{2+} with MnO_4^-.

IV. Iron Ore

A. Iron ore is a mixture of iron oxides and iron sulfides with other minerals. In these experiments we will simulate an ore sample with pure Fe_2O_3.

Put 20 mL of 12 M HCl into a clean 100 or 150 mL beaker. Add about 1 cm^3 of Fe_2O_3. Gently heat the sample, stirring periodically, until the ore is dissolved. If it begins to boil, remove the heat. It is best to heat the sample using a hot plate in a hood. If none are available use a burner flame with the beaker solidly supported on a wire gauze.

CAUTION!! HCl, especially when hot is extremely corrosive to flesh. Handle the hot beaker with tongs or a triply wrapped towel. If no hood is available, cover the beaker with a watch glass to cut down on HCl fumes. The droplets collecting on the watch glass will be highly acidic.

Record your observations.

B. Make a diluted sample of the dissolved ore by adding 1 mL of the warm solution **TO** about 5 mL of water. Add 2 drops of 0.02 M MnO_4^- and record the results. (We are using MnO_4^- here, instead of CrO_7^{2-}, to test for Fe^{2+} because this solution is highly colored.)

Make a second diluted sample of the dissolved ore and add 1 drop of 1 M SCN^-. Record your observations.

What forms of iron ions are present in the dissolved ore? Explain your reasoning.

C. Obtain about 10 mL of 1 M $SnCl_2$. Add this Sn^{2+} a dropperful at a time to the remainder of the dissolved ore sample. Continue until the sample is colorless. The sample should be kept warm during these additions and should be constantly stirred.

Make two 1 mL:5 mL dilutions of the colorless ore solution (see part IV.B). Add 20 drops of $Cr_2O_7^{2-}$ to one dilution; one drop of SCN^- to the other. Record your observations.

What forms of iron ions are present in the ore solution after Sn^{2+} treatment? Write a balanced chemical equation for the reaction of the ionic iron with Sn^{2+} (assume the tin product is Sn^{4+}). Briefly explain your reasoning.

D. Describe how you would prepare a solid ore sample for analysis of total iron content by MnO_4^- titration. Explain why each step in the preparation is necessary.

V. Titration Analysis

A. Obtain about 50 mL of 0.02 M $KMnO_4$ solution in a clean, labeled beaker. Take precautions to not dilute or contaminate your sample of this solution. Prepare a buret and fill it with $KMnO_4$ solution as directed in Appendix D.1. Record the exact concentration of this solution in the table on the next page.

B. Weigh 0.5 to 0.8 grams of $FeSO_4(NH_4)_2SO_4(H_2O)_6$, ferrous ammonium sulfate, into a 250 mL Erlenmeyer flask. The exact mass should be known to at least three significant figures.

Add 100 mL of distilled water, 10 mL of 3 M HCl, then swirl to dissolve the Fe^{2+} salt.

[Note: If the sample was an Fe^{3+} salt, you would now heat it to near boiling, add a couple of drops of 1 M $SnCl_2$ and swirl. You would continue this dropwise addition and mixing until the yellow color of Fe^{3+} was just dispelled.]

C. Add 10 mL of 3 M H_3PO_4 Solution. (The phosphate sharpens the end point by binding to the iron ions.)

D. Record the initial volume reading of the buret, then titrate the Fe^{2+} sample. The end point is reached when the addition of one drop of MnO_4^- causes a pink color that persists in the solution for about 30 seconds. Record the final buret reading and the total volume of MnO_4^- used.

E. If time allows, weigh out another sample of the Fe^{2+} salt and titrate it. Collect titration data from other students.

DATA TABLE

KMnO$_4$ solution = _____ M

Titration Data

grams of Fe^{2+} salt _____ _____ _____ _____ _____

initial buret reading _____ _____ _____ _____ _____

final buret reading _____ _____ _____ _____ _____

mL MnO$_4^-$ solution _____ _____ _____ _____ _____

moles of Fe^{2+} _____ _____ _____ _____ _____

moles of MnO$_4^-$ _____ _____ _____ _____ _____

VI. In part II.C you wrote a balanced chemical equation for the reaction of Fe^{2+} and MnO_4^-. Show mathematically how your titration data confirm or deny this equation. Explain your reasoning. If the equation does not fit the titration data, find a new equation that does fit both the qualitative data and titration data. Explicitly show that this new equation does fit all the data.

PART TWO

You are to apply the chemistry and techniques from Part One to determine the iron content of "unknown" samples. Your instructor will decide the basic objective of the analysis. The following are possibilities.

Alternate no. 1: You will be given a solution, salt or powdered ore unknown. You are to determine the iron concentration of it in M or weight %. You will be judged on the aptness of your procedure, the reproducability of your results and the accuracy of your results. The purpose is to perfect your analysis skills.

Alternate no. 2: You will be given, or are to collect, several natural substances containing iron (raw ore samples, mine tailings, commercial products.) You are to analyze these for iron content noting any analytical difficulties and any trends among related samples. You will be judged on the aptness of your procedure, the reproducability of your results and the explanations you develop. The purpose is to sharpen your analytical skills and to give you experience with "real world" analysis problems.

Your report on this work is to contain the following sequential parts.

1. A brief statement of the scope of the experiment.

2. A brief outline of procedure.

3. Presentations of raw data.

4. Mathematical treatment of the data.

5. Conclusions and explanations.

DISSOLVED OXYGEN—Exp. M-2

Name_____ Lab Section_____

Lab Partner _____

Pre-lab assignment: Study Appendix D.1.E on use of burets.

PART ONE

I. Preparation of Solutions for Analysis.

You are to prepare three solutions using flasks specially designed to exclude air pockets. These will be called DO flasks (Dissolved Oxygen flasks). Your instructor will have demonstrated their use and safety precautions.

A. Secure three of these flasks, rinse them with distilled water, and label them A, B, C.

Also secure, in clean, labeled beakers: about 2g of d-glucose ($C_6H_{12}O_6$), and 15 mL of "microbe water." This latter is a solution containing a high population of bacteria.

B. Fill each DO flask 3/4 full with distilled water. Stopper, and shake the flask. Unstopper, restopper, and then shake again. This is meant to saturate the water with O_2 from the air. This is called aerating the sample.

You will need another 300 mL of aerated distilled water to top off your DO samples. Prepare this water using any appropriate flask.

C. Add the following to the designated DO flask, mix to dissolve.

Flask A: nothing more.

Flask B: about 1g of d-glucose.

Flask C: about 1g of d-glucose and 10 mL of microbe water

D. Fill each DO flask to its brim with areated distilled water. Stopper each over a sink. If air bubbles are trapped in the flask, take out the stopper and add more distilled water, then restopper. Mix each flask by inverting it several times.

Set the DO samples aside for later analysis in part III.

CAUTION!!! Both the KOH and the H_2SO_4 are extremely corrosive. If any gets on your person, wash the area immediately with water. Also, **DO NOT MIX THEM AT FULL STRENGTH**, the reaction is violent.

II. Chemical Reactions

Listed below are chemical equations involved in the analysis of dissolved oxygen (DO) in a sample. Use these to help answer the questions in parts II.B to II.G.

$$I_2 + 2\ S_2O_3{}^{2-} \longrightarrow S_4O_6{}^{2-} + 2\ I^-$$

$$2\ Mn^{2+} + 4\ OH^- + O_2 \longrightarrow 2\ MnO_2(s) + 2\ H_2O$$

$$C_6H_{12}O_6 + 6\ O_2 \longrightarrow 6\ CO_2 + 6\ H_2O$$

$$MnO_2(s) + 2\ I^- + 4\ H^+ \longrightarrow Mn^{2+} + I_2 + 2\ H_2O$$

A. In separate, clean, towel dried, labeled 100 mL beakers obtain about

   1 cm^3 of manganous chloride,
   3 pellets of potassium hydroxide,
   1/2 cm^3 of potassium iodide,
   1 cm^3 of sodium thiosulfate,
   20 mL conc. sulfuric acid (also save for part III).

B. Add 20 mL of water to the manganous chloride and mix. This will be called the reaction beaker. Record your observations below, speculate on what chemical reaction(s) occurred, and briefly explain your reasoning.

Observations:

Chemical Reaction(s):

Reasoning:

C. Add 20 mL of water to the KOH pellets, mix until dissolved. Add this solution to the reaction beaker, mix.

Observations:

Chemical Reaction(s):

Reasoning:

D. Dissolve the KI in 20 mL of water, add this to the reaction beaker, mix.

Observations:

Chemical Reaction(s):

Reasoning:

E. Using a dropper, add about 1/2 mL (10 drops) of conc. H_2SO_4 to the reaction beaker, mix.

Observations:

Chemical Reaction(s):

Reasoning:

F. Dissolve the $Na_2S_2O_3$ in about 20 mL of water. Add this a few milliliters at a time to the reaction beaker; mix between additions.

Observations:

Chemical Reaction(s):

Reasoning:

G. Use the information gained above to explain how you would quantitatively analyze a solution for its dissolved oxygen using a standardized solution of $Na_2S_2O_3$.

(You might describe what chemical treatment you would give the sample before titration, giving reasons for this treatment. Describe the titration. How will you know when you have reached the end-point? Chemically what happens at the end-point?)

III. Analysis of the DO Samples

A. Obtain about 50 mL of 0.02 M $Na_2S_2O_3$ solution in a labeled beaker. Take care to not dilute or contaminate your sample of this solution. Prepare a buret and fill it with the thiosulfate solution as directed in Appendix D.1.E. Record the exact concentration of the solution in the table on the next page.

B. Obtain, in clean, dry, labeled 100 mL beakers,

   10 mL of 2 M Mn^{2+} solution
   10 mL of 10 M KOH, 1 M KI solution
   15 mL of conc. H_2SO_4 (from part II)
   10 mL of starch solution

Dedicate a dropper to each solution. In doing the work below, take care to **NOT MIX UP THE DROPPERS**.

C. Remove the stopper from DO flask A. Add 2 droppersful (about 1 1/2 mL) of 2 M Mn^{2+} without touching the dropper to the flask's solution. Do this by placing the tip of the dropper just below the rim of the flask and letting the Mn^{2+} solution run down the inside of the rim.

Over a sink, restopper the DO flask, taking care that no air bubbles are trapped inside. If air is trapped, remove the stopper, add distilled water, and restopper.

Mix the sample several times by inversion.

D. In like manner, add one dropperful of 10 M KOH, 1 M KI solution to the DO flask. Stopper over a sink, and mix by inversion. If any of the excluded solution gets on your hands, wash that area with copious amounts of water.

Let the solution sit for two minutes to allow the precipitate to settle.

E. Invert the DO flask to redisperse the precipitate. Remove the stopper and add 2 droppersful of concentrated H_2SO_4. Restopper and mix as before. Again, wash your hands if excluded solution gets on them.

Let the solution sit for a few minutes to clarify as much as it can.

F. Using a graduated cylinder, carefully measure 100 mL of DO solution A into a clean, distilled water rinsed 250 mL Erlenmeyer flask. Record the initial reading of the buret, then begin titrating the DO sample.

NOTE: The titration should only take a few milliliters of thiosulfate solution, so make small additions even at the beginning.

When the color of the solution has changed from a bright orange to a pale orange, add 2 droppers of starch solution. The remaining I_2 will complex with the starch, giving a deep blue color. Continue to titrate until the solution color changes to colorless or very pale blue. Record the final buret reading in the table.

G. Titrate DO samples B and C using the procedure given in parts II.C to II.F above. Record these data in the table.

H. Collect another set of data from other students and enter these in the table.

DATA TABLE

Concentration of $Na_2S_2O_3$ = _____ M

Titration Data

Sample:	A	B	C	A'	B'	C'
Contents:	_____	_____	_____	_____	_____	_____
	_____	_____	_____	_____	_____	_____
	_____	_____	_____	_____	_____	_____
mL sample titrated	_____	_____	_____	_____	_____	_____
initial buret reading	_____	_____	_____	_____	_____	_____
final buret reading	_____	_____	_____	_____	_____	_____
DO (mg O_2/L)	_____	_____	_____	_____	_____	_____

I. Calculate the concentration of dissolved oxygen (DO) in milligrams of O_2 per liter for each sample. You will need to use the mole relationships discovered in part I.

Enter these numbers in the table. Show your calculation for your Sample ''A'' below.

IV. A. Water saturated with air at 25°C has an O_2 concentration of about 8 mg/L. Compare your results to this number. Offer explanations for why your samples agree or disagree with this number.

B. A standard analysis is to put aerated water, a pollutant and bacteria in a DO flask and then analyze it 5 days later by titration as you did. The drop in DO over that time is called the Biochemical Oxygen Demand (BOD) of the pollutant. Explain why this is an appropriate label.

(You might speculate on what is happening chemically in the sample, and on what role the bacteria play.)

PART TWO

You are to apply the chemistry and techniques from Part One in experiments of your own design. Possible ideas are listed below. Because of time and equipment restraints, your instructor may organize investigative teams or limit your options. Write up your Reports as described for on page 224 for other Open Inquiry Experiments.

NOTE: The American Public Health Association, in "Standard Methods for the Analysis of Waters and Wastewaters," has established guidelines for BOD analysis. These use natural bacterial waters (i.e., lake waters) and require five or more days for complete reaction. To get the reactions to occur in the few hours of the student lab period, your instructor may have you use hyperactivated colonies of bacteria.

System 1

Investigate the DO level of various water samples. Examples might include fast flowing streams, still deep lakes, water from a nonareated aquarium, water that has had N_2 bubbled through it.

System 2

Determine the experimental BOD of d-glucose and compare it to the theoretical value.

System 3

Investigate the effect of bacteria concentration on the rate of DO depletion.

System 4

Investigate the effect of nutrient (pollutant) concentration on the rate of DO depletion.

System 5

Compare the rates of DO depletion with bacteria from different sources. Possibilities include lake water, "grey water," bakers yeast, pre-incubated versus not incubated (hyperactivated versus natural).

Section Two

OPEN INQUIRY EXPERIMENTS

A. INTRODUCTION TO THE STUDENT

The experiments in this section of the laboratory manual are called Open Inquiry experiments. With this section you will be designing and carrying out your own experiments much as a research scientist must do. One way to look at these Open Inquiry experiments is that they are designed to give you the time and equipment to "do your own thing," to do experiments that make sense to you and that will help you personally to learn some chemistry.

The laboratory manual contains only lists of what are called systems for each general area (i.e., mass relationships, heats, etc.). These systems are only suggestions to help you get started. You may work on one or more of these systems or invent your own. Remember, you may do any investigation that you want. The only limitations are: (1) that the investigation should be connected to the content of the Guided Inquiry (Section One) experiment previously done, (2) that the equipment and chemicals necessary for the investigation are available or easily constructed (see your lab instructor), and (3) that your safety and the safety of others is safeguarded (make sure your laboratory instructor knows what you are doing).

There are no detailed instructions on how to approach these systems. This will be your job. There are many different ways to approach each system. The choice is yours. Following is a list of ground rules and suggestions that you should observe:

(1) You must complete the assigned Guided Inquiry experiment (Section One) laboratory report before beginning your Open Ended experiments.

(2) If you have no ideas of your own on what problem to investigate, read through the systems for that general area. If you are still undecided as to what to do, select a couple of the systems and do some short, qualitative investigations of them. From your observations you should find some aspect of a system to investigate in more detail.

(3) Before beginning your detailed investigation it would be helpful to make a rough outline of the procedures you will use, i.e., what chemicals and equipment you will need, what you will measure, what you will vary, what you will hold constant. **Check with your lab instructor to be sure your experiments are reasonably safe.**

(4) You should spend at least 2/3 of the lab time available in actual experimental work. If you finish an investigation to your satisfaction and there is still lab time remaining, start investigating another system.

Laboratory Report

The laboratory report on your Open Inquiry experiments should be written so that any person could follow it. Try to be both concise and clear. The report should be divided into the four parts listed below.

I. **Introduction**: This should be just a few sentences stating the problem or describing the system you investigated. If you had any predictions as to what would happen include them here.

II. **Procedures**: This should be a brief outline of the experimental procedures you used including the equipment and chemicals employed and what variables in the experiment were measured. A detailed step-by-step account of the experiments is not required.

III. **Data Presentation and Analysis**: Data should be presented in organized tabular form and labeled so it is easy for a reader to follow. Algebraic equations, mathematical relationships, and graphical representation of the data should be used when possible and appropriate. All of these should be developed and presented in a way so that their connection and relevance to the data are clear.

IV. **Findings**: This part should include your conclusions, explanations and any generalizations you feel you can make about the chemistry. The logic you used in arriving at these ideas should be clearly stated. Try to also include logical alternative explanations and hypothesis.

Your grade will be derived from how you investigated your system, how thoroughly you investigated the system in the time allowed, how the lab report and data are presented, and the reasonableness of your data analysis and explanations. It is especially important that all of your conclusions are logically consistent with the data you collected. Although you are encouraged to work with your partner, and to share ideas and information with others, you will be expected to write up your final report independently.

Michael R. Abraham
Michael J. Pavelich

B. MEASUREMENT SYSTEMS

System 1

Investigate the uncertainty of measurement instruments. Possibilities include:
    graduated cylinders
    meter sticks
    thermometers

System 2

Investigate the uncertainty involved in determining the volume of a regularly shaped solid.

System 3

Investigate the uncertainty of measuring an object with a single instrument as compared with using several different examples of the instrument.

System 4

Investigate the physical relationships of metal cylinders made of metals other than brass or aluminum. Compare their physical relationships with brass and aluminum.

System 5

Investigate the graphical relationship between the temperature of a cooling liquid and the time passed.

System 6

Investigate, by graphing, the relationship between two variables looked up in a handbook. Possibilities include:
    density and melting point of materials (metals)
    density and heat capacity of materials

System 7

Investigate the volume of water and its height in various containers (i.e., funnel or flask).

System 8

Investigate the measurement relationships associated with any other system or any modification of any of the above systems. For safety reasons, discuss your system with the instructor before proceeding.

C. MASS RELATIONSHIPS SYSTEMS

System 1

Investigate the mass relationships associated with heating various compounds. Possibilities include:
  Magnesium—Mg
  Boric Acid—H_3BO_3
  Sodium Bicarbonate—$NaHCO_3$
  Copper (II) Carbonate—$CuCO_3$

System 2

Investigate the mass relationship associated with heating hydrated chemical substances. One or more of the following are possibilities:
  Hydrated Ferrous Sulfate
  Hydrated Barium Chloride
  Hydrated Cobaltous Chloride
  Hydrated Copper (II) Chloride

System 3

Investigate the mass relationships associated with adding 4 M hydrochloric acid (HCl) dropwise to sodium carbonate (Na_2CO_3) or to sodium bicarbonate ($NaHCO_3$). **CAUTION**: See System 6.

System 4

Investigate the mass relationships associated with heating a mixture of sulfur and a metal. (**CAUTION**: obnoxious odor may form. Use a hood or an aspirator-funnel arrangement). Possible metals are:
  Copper
  Iron
  Lead

System 5

Investigate the mass relationships associated with the formation of precipitates when a solution of one ionic compound is mixed with a solution of another ionic compound. Possibilities include:
  Lead (II) Nitrate and Potassium Iodide
  Barium Chloride and Sodium Sulfate
  Calcium Chloride and Sodium Carbonate
  Cobalt (II) Nitrate and Sodium Phosphate
  Nickel (II) Sulfate and Potassium Hydroxide

System 6

Investigate the mass relationships associated with the addition of metals to diluted solutions of hydrochloric acid (HCl) or sulfuric acid (H_2SO_4). (Hints: dilute concentrated acids by adding acid to water. The strength of acid solutions can be determined by titrating with a standardized NaOH solution). **CAUTION**: HCl, H_2SO_4 and NaOH are corrosive to the skin, wash spills with large amounts of water. Possible metals include:

  Iron
  Zinc
  Magnesium
  Aluminum

System 7

Investigate the mass relationships associated with any other system or any modification of any of the above systems. For safety reasons, discuss your system with the instructor before proceeding.

D. HEAT LAWS SYSTEMS

System 1

Investigate the heat lost or gained when a specific chemical or group of chemicals are added to water. The following are possibilities:

    sodium nitrate
    sodium chloride
    hydrated calcium chloride
    ammonium nitrate
    sodium hydroxide (**CAUTION**: This material is corrosive and should not be allowed to come in
                    contact with your skin.)

System 2

Compare the heat lost or gained by hydrated versus anhydrous compounds when they are dissolved in water. It may be necessary to make your own anhydrous salts from the available hydrate. Possibilities include:

    $MgSO_4(H_2O)_7$ vs. $MgSO_4$
    $CaCl_2(H_2O)_2$ vs. $CaCl_2$
    $CuSO_4(H_2O)_5$ vs. $CuSO_4$
    $CoCl_2(H_2O)_6$ vs. $CoCl_2$

System 3

Investigate the heat gained or lost when various bases are mixed with various acid solutions. For safety reasons use acid concentrations of 2.0 M or less. The bases can be used in solid form or in solution. Possible chemicals include:

Bases	Acids
NaOH	HCl
KOH	H_2SO_4
$Ca(OH)_2$	HNO_3

System 4

Investigate the effect of changing the calorimeter conditions on the heat of a solution reaction. See Systems 1-3 for possible reactions. Calorimeter variables include:

    glass vs. metal vs. polystyrene container
    open vs. closed calorimeter
    amount of water
    initial temperature of water
    ice present in the water
    thickness of the polystyrene walls

System 5

Investigate the heat lost or gained when a ''hot'' or ''cold'' piece of metal is added to water. The following are possibilities:

Aluminum
Zinc
Brass

System 6

Investigate the temperature changes occurring during a phase change. The following are possibilities:

Melting of ice
Melting of wax

System 7

Design and carry out experiments to measure the heats of combustion of organic compounds. **Caution: Have your instructor check your set-up for safety.** The following are possibilities:

methanol
ethanol
propanol
paraffin

System 8

Investigate the temperature of the air and other gases above water as the water is heated to boiling.

System 9

Investigate the heat lost or gained with any other system or any modification of any of the above systems. For safety reasons, discuss your system with the lab instructor before proceeding.

E. GAS SYSTEMS

System 1

Investigate the behavior of gases other than air. Possibilities include oxygen (can be generated by heating potassium chlorate with a pinch of manganese dioxide as a catalyst), or carbon dioxide (can be generated by adding dilute acid to calcium carbonate). **CAUTION!!** When diluting acids extreme care must be taken not to allow the acid to come into contact with your skin or clothes. Also add acid to water **not vice versa** when diluting.

System 2

Investigate the relationship between the volume and temperature of a gas at constant pressure.

System 3

Investigate the relationship between the amount of gas and its pressure at constant temperature and volume.

System 4

Investigate and compare the vapor pressures of liquids using a 250 mL suction flask instead of a 125 mL flask.

System 5

Investigate and compare the vapor pressure of pure liquids with solutions of those liquids containing measured amounts of dissolved solid chemicals.

System 6

Investigate any other gas system or investigate a modification of any of the above systems. For safety reasons, discuss your system with the lab instructor before proceeding.

F. PERIODICITY SYSTEMS

System 1

Investigate the ionic nature of a series of substances. (Hint: the ionic nature of a substance can be related to its electrical conductivity. This property can be measured qualitatively with a ''light bulb'' apparatus, and quantitatively with a conductance meter. See your laboratory instructor for instructions on their use.)

a. To be measured qualitatively:

H_2O (distilled)	magnesium chloride, $MgCl_2$
sodium chloride, NaCl	sugar, CH_2O
sodium bromide, NaBr	potassium chloride, KCl
cyclohexane, CH_2	potassium bromide, KBr
potassium chloride, KCl	

b. To be measured quantitatively (Hint: measure the resistance of very weak solutions; no more than 0.0025 moles of solid in 250 mL of solution.):

1. Nitrate salts of Li^+, Na^+, K^+, Mg^{2+}, Ca^{2+}, Sr^{2+}, Ba^{2+}, Al^{3+}

2. Sodium salts of F^-, Cl^-, Br^-, I^-

System 2

Investigate the extent of the solubility of a series of soluble salts. One or more of the following series are possible.

a. NaF, NaCl, NaBr, NaI
b. KF, KCl, KBr, KI
c. LiCl, NaCl, KCl
d. LiBr, NaBr, KBr

System 3

Investigate the reactions between solutions of a series of cations with solutions of an anion. A possible series of cations include:

Mg^{2+}, Ca^{2+}, Sr^{2+}, Ba^{2+} as a water soluble nitrate, chloride, etc.

One or more of the following are possible anions.

$CO_3^{2-}[(NH_4)_2CO_3]$
SO_4^{2-} (dilute H_2SO_4)
OH^- (as NaOH and as NH_4OH)

System 4

Investigate the chemical properties of any other series of elements or ions or investigate a modification of any of the above systems. For safety reasons, discuss your system with the lab instructor before proceeding.

G. ACID AND BASE SYSTEMS

System 1

Investigate the pH properties of acids and their "conjugates." Below are possible acids to investigate:

acid	conjugate
$HC_2H_3O_2$	$NaC_2H_3O_2$
HCl	$NaCl$
$H_2C_2O_4$	$Na_2C_2O_4$
H_2SO_4	Na_2SO_4
H_3PO_4	Na_3PO_4

System 2

Investigate the neutralization of an oxalic acid solution

$$H - O - \underset{\underset{O}{\|}}{C} - \underset{\underset{O}{\|}}{C} - O - H \text{ (s)} \quad \xrightarrow[H_2O]{in} \quad 2H^+(aq) + {}^-O - \underset{\underset{O}{\|}}{C} - \underset{\underset{O}{\|}}{C} - O^-(aq)$$

when it reacts with a standardized NaOH solution.

System 3

Investigate the pH properties of basic compounds when they interact with the strong acid HCl in solution. Possible bases are:

Na_2HPO_4

Na_3PO_4

$NaOH$

Na_2CO_3

NH_4OH

System 4

Investigate the pH at which different indicators change color. (Hint: titrate 25 mL of a 0.1 M acetic acid solution with a standardized base and/or titrate 25 mL of a 0.1 M Na_2CO_3 with 0.1 M HCl.) Below are possible indicators to use:

phenolphthalein   methyl orange

bromocresol green   indigo carmine

alizarin yellow   methyl red

bromthymol blue

System 5

Investigate the acid-base properties of some common household products.

  a. Investigate foods and household chemicals like orange juice, 7-Up, (fresh and "flat"), vinegar, ammonia, etc.
  b. Investigate the power of various commercial antacids.
  c. Investigate the power of various brands of aspirin (Hint: the active drug in aspirin is acetylsalicylic acid.)

System 6

Investigate the pH and conductivity properties of a set of acids or bases. (Hint: for best results solutions should not exceed 0.01 M). Possible acids and bases are:

acids	bases
HCl	NaOH
H_2SO_4	NH_3
H_3PO_4	Na_3PO_4
$HC_2H_3O_2$	$NaC_2H_3O_2$
$H_2C_2O_4$	$Ca(OH)_2$
KHP	

System 7

Investigate the acid-base properties of any other system or investigate a modification of any of the above systems. For safety reasons, discuss your system with the lab instructor before proceeding.

I. EQUILIBRIUM SYSTEMS

System 1

Investigate the solubility of lead salts such as

Lead (II) chloride — $PbCl_2$
Lead (II) bromide — $PbBr_2$
Lead (II) oxalate — PbC_2O_4

These can be made by reacting $Pb(NO_3)_2$ solutions with solutions of NaCl, NaBr or $Na_2C_2O_4$.

System 2

Investigate how I_2 distributes itself between water and C_6H_{12} (cyclohexane) when the two immiscible (insoluble) liquids and I_2 are mixed together.

NOTE: The concentration of I_2 in each liquid can be determined by titrating with a sodium thiosulfate ($Na_2S_2O_3$) solution of known concentration. The titration (analysis) reaction is:

$$I_2 + 2\ S_2O_3{}^{2-}(aq) \longrightarrow 2I^-(aq) + S_4O_6{}^{2-}(aq)$$

System 3

Investigate the "completeness" of precipitation reactions. The following are possibilities:

$Co(NO_3)_2$ and Na_2CO_3
$NiSO_4$ and Na_2CO_3
$Co(NO_3)_2$ and $Na_2C_2O_4$

System 4

Investigate the pH properties of acids. Possibilities include:

Potassium acid phthalate — $HKC_8H_4O_4$
Acetic acid — $HC_2H_3O_2$
Oxalic acid — $H_2C_2O_4 \cdot 2H_2O$
Hydrochloric acid — HCl

System 5

Investigate the pH properties of bases. Possibilities include:

Sodium acetate — $NaC_2H_3O_2$
Sodium carbonate — Na_2CO_3
Potassium hydroxide — KOH
Sodium phosphate — Na_3PO_4
Sodium monohydrogen phosphate — Na_2HPO_4
Sodium oxalate — $Na_2C_2O_4$
Sodium chloride — $NaCl$

System 6

Investigate the effect on pH of mixing an acid with a base. See Systems 4 and 5 for suggestions.

System 7

Investigate the pH and conductivity properties of a set of acids or bases. (Hint: for best results solutions should not exceed 0.01 M). Possible acids and bases are:

acids	bases
HCl	NaOH
H_2SO_4	NH_3
H_3PO_4	Na_3PO_4
$HC_2H_3O_2$	$NaC_2H_3O_2$
$H_2C_2O_4$	$Ca(OH)_2$
KHP	

System 8

Investigate the concentration properties of any other system or modification of any of the above systems. For safety reasons, discuss your ideas with the lab instructor before proceeding.

J. ELECTROCHEMICAL SYSTEMS

System 1

Investigate the effect of decreasing the concentration of one of the 1/2 cells of the $Cu/Cu^{2+} \mid\mid Cu^{2+}/Cu$ cell.

System 2

Investigate the properties of other cell combinations. Possible half cells include: (Note that some 1/2 cells contain more than one dissolved species.)

$$Mg/Mg^{2+} \mid\mid$$
$$Fe/Fe^{2+} \mid\mid$$
$$Pt/Mn^{2+}, MnO_4^-, H^+ \mid\mid$$
$$Pt/Cr^{3+}, Cr_2O_7^{2-}, H^+ \mid\mid$$
$$Ni/Ni^{2+} \mid\mid$$
$$Sn/Sn^{2+} \mid\mid$$

System 3

Investigate the effect of pH on the voltage of various cells using $Pt/CrO_4^{2-}, Cr_2O_7^{2-} \mid\mid$ as a 1/2 cell, with other 1/2 cells.

System 4

Investigate the effect of amount of added $NH_3(aq)$ on the voltage of various cells using Ag/Ag^+ or Cu/Cu^{2+} as a 1/2 cell with other 1/2 cells.

System 5

Investigate the use of other metals besides Cu in the electrolysis described on page 167. Possibilities include Cd and Pb.

System 6

Investigate the electrolysis of ionic salt solutions using the following apparatus. Possible salts to investigate include:

    $CuCl_2$
    KI

System 7

Investigate the electrolysis of a metal in a metal ion solution using the following apparatus. Possibilities include Cu and Pb.

System 8

Investigate the electrolysis of a voltaic cell. Use the following apparatus. See System 2 for half cells.

System 9

Investigate the electrochemical properties of any other system or any modification of any of the above systems. For safety reasons, discuss your system with the lab instructor before proceeding.

K. KINETICS SYSTEMS

System 1

Investigate the kinetic behavior of a "clock" reaction. Possibilities include:

$$IO_3^- + 3HSO_3^- \longrightarrow 3HSO_4^- + I^-$$ (use dilutions of stock solutions "A" and "B")

$$HCHO + SO_3^{2-} + H_2O \longrightarrow CH_2(OH)SO_3^- + OH^-$$ (use dilutions of stock solutions "C" and "D" with phenolphthalein indicator)

System 2

Investigate the effect $[H^+]$ has on the reaction:

$$Zn + 2H^+ \longrightarrow Zn^{2+} + H_2(g)$$

System 3

Investigate the kinetic behavior of a reaction between oxalic acid and potassium permanganate with and without Mn^{2+}. (Hint: start with 5 mL of .755 M oxalic acid and 1 mL .130M $KMnO_4$ and 6 mL H_2O.)

System 4

Investigate the effect of $[H^+]$, temperature, or metal surface area has on the reaction:

$$Fe + 2H^+ \longrightarrow Fe^{2+} + H_2$$

(You might use $KMnO_4$ solution as an indicator:

$$5Fe^{2+} + MnO_4^- + 8H^+ \longrightarrow 5Fe^{3+} + Mn^{2+} + 4H_2O)$$

System 5

Investigate the effect $[I^-]$ (or $CuSO_4$(aq) or $FeSO_4$(aq) or MnO_2(s)) has on the reaction:

$$2H_2O_2 \longrightarrow 2H_2O + O_2(g)$$

(Hint: start with for example, 10 mL of stock solutions of 0.1 M KI, 5 mL of 3% H_2O_2 and 10 mL H_2O.)

System 6

Investigate the kinetic behavior of any other system or any modification of any of the above systems. For safety reasons, discuss your system with the lab instructor before proceeding.

APPENDICES

APPENDIX A

SAFETY IN THE LABORATORY

Any chemical laboratory by its very nature has several potential hazards which can cause injuries such as chemical burns, fire burns, poisoning and cuts. However, serious injuries **rarely** occur in student labs when safety rules such as the following are thoughtfully and strictly observed.

SAFETY PRECAUTIONS

1. **Wear Safety Goggles at All Times**. Injury to the eyes can easily cause a lifetime of blindness. Safety goggles such as those pictured offer excellent protection against such injury. These must be worn over the eyes (not on your forehead!) at all times that **anyone** in the lab has an experiment going.

2. **Wear Proper Clothing**. To protect the feet and legs from chemical spills, wear shoes (not sandals) and full length loose fitting pants in the lab. Also it is advised that old clothes be worn as minor chemical spotting will occur from time-to-time. Tie back long hair so that it does not fall into flames or chemicals.

3. **Do Not Touch, Taste or Smell Lab Chemicals**. Use the proper equipment (spatulas, pipets, beakers— see Appendix B) to handle chemicals. Tasting of lab chemicals is strictly forbidden as is eating or drinking normal foods in the lab. Any of these can easily lead to poisoning. If you must smell a chemical use an indirect wafting technique.

4. **Do Not Use Cracked or Chipped Glassware**. Replace such glassware or fire polish chips at the direction of your instructor.

5. **Dispose of Chemicals as Directed**. Do not put lab chemicals into waste baskets. Waste chemicals should be placed in crocks or disposal containers as directed by the instructor. (Occasionally, the instructor may allow you to wash soluble, harmless chemicals down the drain.)

6. **Know the Location and Use of Safety Equipment**. Your lab will be equipped with safety equipment such as those pictured on the following page. Talk to your instructor about how and when they are to be used. Find at least two ways to exit the lab.

7. **Never Leave a Burner Flame or Reaction Unattended**.

8. **Follow Safety Instructions for Your Specific Experiment to the Letter**. Read Appendix B and other Appendices as needed. Discuss the procedures with the instructor if they are unclear to you.

Shower

Goggles

Eyewash Fountain

Water

CO_2 Fire Extinguisher

WHAT TO DO IN CASE OF ACCIDENTS

Minor Accidents

1. **Skin Contact with Chemicals**. This is the most common form of accident, usually affecting the hands and forearms. The key word for treatment is **WATER**, large volumes of it. **Immediately** wash the affected area with large volumes of water. Do this even if no sensation of burning is felt since many chemicals are slow acting. Use a faucet for the arms and hands.

   For the face, use an eye wash fountain. Keep your goggles in place for the beginning of the wash so as not to wash the chemical into your eyes. For the torso or legs go to the restroom, disrobe as needed and wash the affected area with large amounts of water. Also rinse the affected clothing before using it again. **If an immediate, heavy burning sensation is felt on the torso or legs use the safety shower**. Report the accident to the instructor.

2. **Chemical Spills**. Any spills, no matter how small, on the bench or floor must be taken care of immediately to prevent others from unknowingly getting injured by them. Check with the instructor for clean-up procedures.

   Usually a spill can be handled by wiping it up with a wet cloth or sponge. The spot should then be washed twice more with fresh water and wiped dry.

3. **Heat Burns**. These come from the momentary touching of hot pieces of metal or glass. Hold the burned area under a stream of cold water or pack it with ice. Report the injury to the instructor.

4. **Cuts**. For small cuts, wash them with water and secure a dressing from the instructor.

5. **Small Fires**. These can usually be extinguished by smothering with a double thickness of toweling.

6. **Think, then Act**. Most injuries occur from reacting too fast to a situation; i.e., an overflowing reaction, a falling beaker. (Trying to catch a beaker of corrosive chemical is tricky to say the least.) The best thing to do in such a situation is to shout a warning, back away and let the accident happen. Then think a few seconds about what you can safely do to rectify the situation and do it. Normally, this will simply be to get the instructor.

Serious Accidents

You be the judge of what is serious or not. Remember it is better to be slightly embarrassed than to let a situation worsen. Your immediate reaction should be to:

1. Back away from the danger area.

2. Shout for help.

3. Try not to panic, stay as calm as possible. The instructor will know what to do.

If you are not involved in the accident you may be called on to lend your instructor some assistance. Primarily this assistance will take the form of getting a person under the safety shower to extinguish flames or to wash off a serious spill.

APPENDIX B

COMMON PROCEDURES AND EQUIPMENT

I. BALANCES

The weighing (or massing) of chemicals is one of the most used and potentially most precise measurements employed by chemists. The measurement is done by **counter balancing** the substance of unknown mass against objects of known mass. This balancing act is accomplished by means of a delicate pivot point called a knife edge. Thus the measuring instruments used are called **balances**.

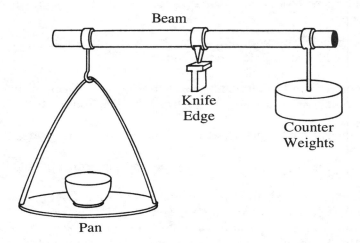

Schematic of a Balance

Three types of balances are pictured and discussed here; the triple beam balance, the top loading electronic balance and the analytical electronic balance. Your instructor will give you specific instructions for the type of balance you will use; however, it is strongly suggested that you study the comments below before doing any balance work.

A. CARE OF THE BALANCE

Because balances are very delicate and fairly expensive instruments, the following rules must be followed.

1. Do not attempt to repair a balance that is malfunctioning. Inform your instructor about the problem.

2. Never place chemicals directly on the balance pans as this will lead to corrosion and eventual malfunctioning. Beakers, paper "boats," etc. should be used to hold the chemical being weighed.

3. Never remove chemicals from a container while it is on the balance pan. This might put too much pressure on the pan. Always remove the container completely from the balance before removing chemicals.

4. Add chemicals to a container on a balance pan carefully, to prevent spills. If a spill does occur, use a balance brush to gently sweep it away. (Be sure the electronic balance is shut off before you start cleaning up.) If the spill involves more than a few grains of material or involves a highly corrosive chemical, see your instructor for clean-up instructions.

5. Do not move a balance. This will upset its leveling and can cause the machinery to jam.

Analytical Electronic Balance

Top-loading Electronic Balance

Triple Beam Balance

B. USE OF THE ANALYTICAL ELECTRONIC BALANCE

(The analytical balance discussed below is the older, partially mechanical version still being used in many laboratories.)

This machine **looks** totally electronic and fool-proof because all that shows is a balance pan, some knobs and a weight read-out window. However, it is basically like other balances; the knife-edge, balance beam and counterweights are just enclosed in the housing. The balance is very precise ($\pm$ .001 to $\pm$ .0002g) making it also very delicate and very easily damaged. Thus it must be used thoughtfully and gently.

Your instructor will explain the use of your specific balance. In general it has four types of controls:

1. Zeroing knob

2. Large Weight knobs—turning these causes weights to be placed on or off the balance beam.

3. Fine Weight knob—this allows for a better estimate of mg weights.

4. On-off lever—this has three positions; Off or Full Arrest (balance beam held in place), Partial Arrest (balance beam has limited movement), Full On (balance beam has complete freedom of movement).

ZEROING

1. Check that the instrument is Off. Adjust all weight knobs to zero and close the doors.

2. Turn to Full On and adjust to read 0.0000g by gently turning the Zeroing knob.

3. Turn to Off.

WEIGHING

1. Place the container to be weighed on the pan. The instrument must be **Off** whenever placing things on or removing them from the pan.

2. Turn lever to Partial Arrest position.

3. Gently and slowly adjust 10 g knob until the projected scale moves abruptly upwards. Then turn the knob back one step.

4. Repeat #3 using the 1g knob and then the 0.1g knob if there is one on your instrument.

5. Turn the lever gently to the Off position and then to the Full On position. Be sure the doors are closed.

6. When the projected scale stops moving, adjust the Fine Weight knob as directed by your instructor.

7. Read and record the mass. (Some machines use a comma as a decimal point in the gram weight.) Be sure your final weight is taken with the lever in the Full On, **not the Partial Arrest** position.

8. Turn the instrument Off.

C. USE OF THE TOP LOADING ELECTRONIC BALANCE

(The balance discussed below is the newest, totally electronic version.)

This instrument uses a pressure sensitive strain gauge to determine mass. The mass of the object affects the electronic signal passing through the gauge; this signal gets displayed as a digital readout of grams. The instrument can be as precise as many analytical balances (± 0.001 g) and is easier to use. Such top loading balances are becoming the instrument of choice for routine lab work. However, they are delicate and expensive to repair. Thus they must also be used thoughtfully and carefully.

The controls will be explained by your instructor but, in general, they are as follows:

1. On-Off switch

2. Calibration switch

3. Tare switch

WEIGHING

1. Turn the instrument on using the On-Off switch.

2. Push the Calibration switch. The readout will rapidly run through a series of displays as the electronics are calibrated. When the display settles on 0.00 g, the calibration is complete.

3. Place the object to be weighed on the pan, wait a few seconds for the display to settle down, then read the gram mass of the object.

4. Turn the instrument Off when you are through with your weighings.

TARING

The Tare switch allows you to reset the display to 0.000 g even with an object on the balance pan. It is used when you ONLY need the mass of the chemical and not the mass of the chemical and container. Use the Tare feature only when you are absolutely sure you will never need to know the mass of the container. If you are unsure, ignore the Tare switch, and record all masses.

1. Weigh the container as described above. After the display settles down, push the Tare switch. The display should now read 0.000 g.

2. Add the desired chemical to the container. The display now show the weight of only the added chemical.

D. USE OF THE TRIPLE BEAM BALANCE

With this instrument, all the essential parts of the balance are visible—the knife edge, balance beam and counter weights. It can give a precision of ± 0.02g if handled properly. Be careful not to move the balance arm sideways (by bumping it) as this might move the knife edge out of position.

ZEROING

1. With nothing on the pan, set all counterweights to their zero position. Be sure your fingers are clean and dry before you handle the weights. It is best to use a spatula or a pencil eraser to move the 0.1g sliding counter weight.

2. When zeroed, the pointer will swing an equal distance on either side of the center line or it will rest at the center line. If this is not the case, adjust the screw nuts at the extreme left hand side of the balance.

WEIGHING

1. Place the container to be weighed on the pan. Using the instructions given in #1 above, move the counterweights until the pointer again rests on or makes equal swings around the center line. Be sure the larger weights are properly seated in the notches on the arms. Read and record the total mass then return the weights to their zero positions.

E. TIPS ON PRECISE WEIGHING

To obtain the most precise (reproducible) results from your weighing, the following procedures should be observed. These are particularly important when using the high precision analytical balance.

1. Check the leveling bulb on the instrument. If it is not dead center, have your instructor make the needed adjustments.

2. Always use the same balance in weighing for a particular experiment. This will allow any small constant error in the balance to be canceled out in your final calculations.

3. Recheck the zero adjustment of the balance occasionally.

4. Be sure the balance and the area around it are kept spotlessly clean.

5. Never use your fingers to directly handle a container you are weighing. Finger prints weigh a few mg each and slowly evaporate. Handle your containers with tongs or with a paper band as shown.

Handling a Weighing Bottle with a Paper Band

Transfer Procedure

II. USING STOCK REAGENT BOTTLES

All the chemicals you will use will be available in stock reagent bottles on the laboratory reagent shelf. The standard procedures for getting what you need from these bottles are described below. These procedures are to be used for two reasons. (1) To prevent contamination of the chemicals. Impure chemicals can give improper experimental results. (2) To prevent excessive wasting of chemicals. Most of the reagents are quite expensive thus overuse of them is economically unsound.

In dealing with solid chemicals be sure the contents of the bottle are loose and in small pieces before you make a transfer. Often the chemical can be loosened by tapping the **closed** bottle against the palm of your hand. If this doesn't work, consult your instructor. Three methods of transferring solids are shown in the figures. The "hollow stopper" method and the direct pouring method are preferred. If you must use a spatula to scoop out the chemical be sure it is clean and dry. Do not go from one reagent bottle to the next without thoroughly cleaning and drying the spatula.

The Hollow Stopper Method

Direct Transfer

May Lead to Contamination

Transferring liquids is much simpler because of their fluidity. The preferred technique is shown in the figure, gentle pouring with the lip of the reagent bottle resting on the lip of the beaker. This technique prevents splashing and allows you to control the amount transferred. When you are pouring, hold onto the beaker to prevent it from tipping. If you need both hands to control the reagent bottle, have your partner hold the beaker. To prevent spills always use a receptacle whose opening is much larger than that of the reagent bottle. Thus you should not pour liquids from a reagent bottle directly into a test tube, nor should you pour from a one gallon reagent bottle into a 50 ml beaker.

Liquid Transfer

Other rules that should be followed are:

1. Wipe up any spills immediately. If liquid runs down the side of a reagent bottle, wipe the bottle with a damp cloth.

2. Do not contaminate the reagent bottle lids. Handle the lids only by the finger grips provided. If the lid has a flat top, set it **upside-down** on the bench. If the top is not flat, do not set it on the bench, hold the lid as shown in the figures. Always replace the lid immediately after using the reagent bottle.

3. Take only what you need. A useful guide is to assume 1 gm of chemical is about 1 cm³. One cm³ is about the volume of the exposed eraser on a new pencil. (Also 1 cm³ is about 1/4 the volume of the last joint on the little finger.)

4. Don't return unused chemical to a reagent bottle. Dispose of unused chemicals as directed by your instructor.

5. Never remove a reagent bottle from the reagent shelf. This is discourteous to others who need the chemical.

6. Label the receptacle so that you will know later which chemical it contains. Most glassware have white, etched glass spots for labelling. Pencil marks work best on these since the marks can be easily erased.

III. BURNER HEATING

The principle sources of heat for most experiments are gas burners which use natural gas (primarily methane, CH_4) and oxygen from the air to produce a hot, clean flame. When used properly the burner gives a flame whose temperature approaches 1500°C. One of the simpler types of burners, called a bunsen burner, is shown below.

Bunsen Burner

Two-Coned Flame

The following procedure is suggested for the first-time lighting of a particular burner.

1. Rotate the barrel to close the air holes. Shut the gas valve on the bottom of the burner, then open the valve by 1/4 turn. Connect the tubing to the gas outlet on the bench.

2. Light a match and hold it by the barrel of the burner so that the flame licks over the rim of the barrel. Turn the gas valve of the bench outlet on full. This will produce a "lazy" flame.

3. Open the air holes by rotating the barrel until a sharp, two-coned flame is obtained. Adjust the gas valve on the burner to get a two-coned flame that is 1" to 2" high.

If the settings on the burner are not altered, subsequent lighting of the burner can be done by just holding match flame to the barrel tip and turning on the bench valve.

The sharp, two-coned flame should always be used as this is the most consistent. To heat an object gently, move the flame in and out of position or place the object a few inches above the flame. Do not attempt to use a "lazy flame." For heavy heating, place the object at the tip of the center cone as this is the hottest part of the flame.

Solid chemicals can be burner heated in porcelain containers (such as crucibles or evaporating dishes) or in Pyrex type test tubes. Do not burner heat a solid in a beaker or flask as cracking of the glassware can occur. When heating liquids in a beaker or flask, it is suggested that you insert a stirring rod or boiling beads to prevent large gas bubbles from forming. **DO NOT** dry glassware using the burner. The heating is too intense and uneven and can result in shattering of the glassware. Use a drying oven or a clean towel if your glassware must be dried.

Heating Solids

Heating Liquids

**Do Not Dry Glassware
with a Burner**

IV. MANIPULATION OF GLASS TUBING AND RODS

In some of your experiments you may need to fashion glass tubing into a certain size and shape or you may need to make stirring rods. The following procedures should be followed in such work.

1. BREAKING GLASS TUBING OR RODS

Using a triangular file or a metal scorer, make a single sharp scratch on the tubing at the point where the break is needed. Moisten the scratch. Grasp the tubing with the scratch facing away from you and with the thumbs located behind and on either side of the scratch. Apply a quick outward breaking force with the thumbs at the same time pulling away from the center (see figure below). The glass should break cleanly.

Scratching the Glass

Breaking the Glass

If you have trouble getting clean breaks, make the scratch a bit deeper. Also your results will improve with practice. Small spurs of glass can be smoothed by stroking them with a wire gauze.

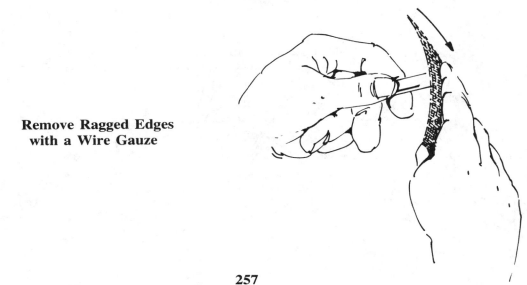

**Remove Ragged Edges
with a Wire Gauze**

2. FIRE-POLISHING

The edge of a cut piece of tubing is very sharp and should be smoothed before use. Place the edge to be polished at the tip of the center cone of a burner flame and rotate the rod to insure even heating. Continue this action until the flame shows a yellow color (caused by Na from the glass being vaporized and ionized). At this point the glass should have softened enough to form a smooth edge. If the edge still appears sharp, continue heating. Avoid excessive fire-polishing as this can lead to a narrowing of the tube opening.

Place the hot piece of glass on a wire gauze to cool. Allow it to cool for 5 to 10 minutes then test the temperature of the heated end by gingerly touching it with a wet finger. **CAUTION: HOT GLASS LOOKS JUST LIKE COLD GLASS**. Severe burns will result from handling uncooled glass.

Fire Polishing

Wire Gauze

Cool for 5 to 10 Minutes

3. BENDING GLASS TUBING

In order to properly bend a piece of tubing, the entire area of the bend must be softened. This is accomplished by using a wing-top (flame spreader) on the burner. Rotate the tubing in the high part of the wing-tip flame until the flame yellows and the glass becomes quite soft. Remove the tubing from the flame and smoothly bend it to the desired angle. (Too quick a bending will cause elongation and distortion of the glass.) Hold the bend in place for 10 to 15 seconds to allow it to harden. Place the bent tubing on a wire gauze to cool as described under FIRE-POLISHING.

Wing-Top

Heating for a Bend

258

Bend Smoothly and Hold Until Solid

4. INSERTING TUBING INTO A STOPPER

Inserting any glass stem (tubing, rods, thermometers) into the hole of a stopper is a potentially dangerous operation. The most common accident is for the glass to snap and be driven into the hands of the experimenter. Thus the task demands some skill and strict attention to safety precautions.

The tube to be inserted should have rounded edges (be fire-polished). The hole in the stopper should be just slightly smaller than the outer diameter of the tubing. Always insert from the side of the stopper that is to have the longest projection of tubing.

Wet the tip of the tubing and the stopper hole with water. Using toweling to protect **both** hands, grasp the tubing with the fingers only, about **one inch** from the end to be inserted. Insert by using a twisting motion with only moderate pushing. Continue this action, periodically lubricating with water, until the desired length of tubing is inserted. NOTES: The toweling will protect both hands from receiving deep cuts if the tube breaks. By holding the tubing near the end being inserted the tension on the rod is minimized. Be careful that the palm of the hand does not push against the glass as this can snap it. If the tubing is not inserting smoothly, **DO NOT FORCE IT**, consult with the instructor.

twisting motion

towel

Inserting into a Stopper

259

OTHER COMMON EQUIPMENT

Beaker

Bottle

Graduated Cylinder

Florence Flask

Erlenmeyer Flask

Watch Glass

Funnel

Crucible

Test Tube Rack

Medicine Dropper

Volumetric Pipet

Glass Rod

Utility Clamp

Pinch Clamp

Iron Ring

Evaporating Dish

Ring Stand

Buret Clamp (double)

Volumetric Flask

Screw Clamp

Filter Flask

Buret

Clay Triangle

Mortar and Pestle

Pipet Bulb

Test Tube Holder

Crucible Tongs

Litmus Paper

Test Tube

Brushes

File

APPENDIX C

TABLES OF DATA

The following tables are necessarily abbreviated and incomplete. Readers are referred to more complete listings such as the Chemical Rubber Company's *Handbook of Chemistry and Physics*.

Table C-1: Physical Constants and Conversion Factors

1 atmosphere pressure (atm) = 760 mm of mercury(torr)

1 ampere current(amp) = 1 coulomb/sec

1 Faraday = 96,500 coulombs, the amount of electricity which will deposit one equivalent of a substance at an electrode

$$R \text{ (gas constant)} = 0.08206 \ \frac{\text{L-atm}}{\text{°K-mole}}$$

$$= 1.987 \ \frac{\text{cal}}{\text{°K-mole}}$$

$$= 8.314 \ \frac{\text{J}}{\text{°K-mole}}$$

$$1 \text{ cal} = 4.184 \text{ J}$$

Table C-2: Identification of Selected Chemicals

Products of chemical reactions can be identified by their chemical or physical properties. Following is a key to identifying some common chemicals.

Formula	Color	Odor	State at 25°C	Solubility in H_2O	Miscellaneous
H_2	none	none	gas	no	burns explosively
O_2	none	none	gas	no	makes glowing split flame
CO_2	none	none	gas	yes	turns lime water milky (saturated $Ca(OH)_2$)
NO_2	brown	choking	gas	yes	solution is acidic
I_2	purple	medicine	solid	yes	purple in hexane (or cyclohexane)
Br_2	red-brown	choking	liquid	yes	orange in hexane (or cyclohexane)
Cl_2	pale green	choking	gas	yes	yellow in hexane (or cyclohexane)
NH_3	none	choking (smelling salts)	gas	yes	solution is basic
SO_2	none	suffocating	gas	yes	solution is acidic
H_2S	none	choking	gas	yes	solution is neutral
HCl	none	choking	gas	yes	solution is acidic
H_2O	none	none	liquid	_____	$CoCl_2$ paper turns pink

Table C-3: Solubility Rules

Following are general rules concerning the solubility of common ionic salts in water. These substances are considered soluble if they can form solutions of concentration greater than 0.1 M. They are considered partially soluble if they form solutions between 0.01 M and 0.1 M. They are considered insoluble if they form solutions with concentrations less than 0.01 M.

Ion	Rule	Important Exceptions
Na^+, K^+, NH_4^+	salts of these ions are soluble	————————
$NO_3^-, ClO_3^-,$ $ClO_4^-, C_2H_3O_2^-$	salts of these ions are soluble	$AgC_2H_3O_2$ partially soluble
Cl^-, Br^-, I^-	salts of these ions are soluble	Ag^+, Pb^{2+}, Hg_2^{2+} salts of these ions are insoluble $PbCl_2$ partially soluble
SO_4^{2-}	salts of this ion are soluble	$BaSO_4, PbSO_4, SrSO_4$ are insoluble $CaSO_4, Hg_2SO_4, Ag_2SO_4$ are partially soluble
$CO_3^{2-}, PO_4^{3-},$ SO_3^{2-}	salts of these ions are insoluble	Na^+, K^+, NH_4^+ salts of these ions are soluble
OH^-	salts of this ion are insoluble	$NaOH, KOH, NH_4OH,$ and $Ba(OH)_2$ are soluble. $Ca(OH)_2$ is partially soluble
S^{2-}	salts of this ion are insoluble	$Na^+, K^+, NH_4^+ Mg^{2+}, Ca^{2+}, Sr^{2+},$ and Ba^{2+} sulfides are soluble

Using these rules it is possible to predict the formation of precipitates in chemical reactions. For example in double displacement reactions (metathesis reactions) the solution of one ionic solid is added to the solution of a second ionic solid. If a precipitate is formed, use of solubility rules allow the identification of the precipitate. For example:

$$Ba(NO_3)_2(aq) + Na_2SO_4(aq) \longrightarrow 2NaNO_3(aq) + BaSO_4(s)$$

nitrates are soluble	Na$^+$ salts are soluble, sulfates are soluble with exceptions	nitrates are soluble, Na$^+$ salts are soluble	sulfates are soluble, Ba^{2+} sulfate is an exception

Table C-4: Selected Acid-Base Indicators

Indicator	pH Range	Color Change(pH)
methyl orange	3.2—4.4	red to yellow
methyl red	4.8—6.0	red to yellow
bromothymol blue	6.0—7.6	colorless to yellow to blue
phenolphthalein	8.2—10.0	colorless to pink
alizarine yellow	10.1—12.0	yellow to red

Table C-5: Strength of Acids in Water at Room Temperature

Strength	Acid	Reaction	K_a*
strong	perchloric acid	$HClO_4 + H_2O \longrightarrow H_3O^+ + ClO_4^-$	large
	hydrochloric acid	$HCl + H_2O \longrightarrow H_3O^+ + Cl^-$	large
	nitric acid	$HNO_3 + H_2O \longrightarrow H_3O^+ + NO_3^-$	large
	sulfuric acid	$H_2SO_4 + H_2O \longrightarrow H_3O^+ + HSO_4^-$	large
	oxalic acid	$H_2C_2O_4 + H_2O \longrightarrow H_3O^+ + HC_2O_4^-$	6.5×10^{-2}
	hydrogen sulfate ion	$HSO_4^- + H_2O \longrightarrow H_3O^+ + SO_4^{2-}$	1.3×10^{-2}
	phosphoric acid	$H_3PO_4 + H_2O \longrightarrow H_3O^+ + H_2PO_4^-$	7.5×10^{-3}
	formic acid	$HCHO_2 + H_2O \longrightarrow H_3O^+ + CHO_2^-$	1.7×10^{-4}
	hydrogen oxalate ion	$HC_2O_4^- + H_2O \longrightarrow H_3O^+ + C_2O_4^{2-}$	6.1×10^{-5}
	ascorbic acid	$HC_6H_7O_6 + H_2O \longrightarrow H_3O^+ + C_6H_7O_6^-$	7.9×10^{-5}
	acetic acid	$HC_2H_3O_2 + H_2O \longrightarrow H_3O^+ + C_2H_3O_2^-$	1.8×10^{-5}
	carbonic acid	$H_2CO_3 + H_2O \longrightarrow H_3O^+ + HCO_3^-$	4.3×10^{-7}
	dihydrogen phosphate ion	$H_2PO_4^- + H_2O \longrightarrow H_3O^+ + HPO_4^{2-}$	6.3×10^{-8}
	boric acid	$H_3BO_3 + H_2O \longrightarrow H_3O^+ + H_2BO_3^-$	7.3×10^{-10}
	ammonium ion	$NH_4^+ + H_2O \longrightarrow H_3O^+ + NH_3$	5.7×10^{-10}
	bicarbonate ion	$HCO_3^- + H_2O \longrightarrow H_3O^+ + CO_3^{2-}$	4.8×10^{-11}
	monohydrogen phosphate ion	$HPO_4^{-2} + H_2O \longrightarrow H_3O^+ + PO_4^{3-}$	4.4×10^{-13}
weak	water	$2H_2O \longrightarrow H_3O^+ + OH^-$	$K_w = [H^+][OH^-]$ $= 1 \times 10^{-14}$

*for $HB + H_2O \longrightarrow H_3O^+ + B^-$, $K_a = \dfrac{[H_3O^+][B^-]}{[HB]}$

Table C-6: Standard Reduction Potentials for Selected Half-Reactions

half-reaction	E°(volts)
$2e^- + F_2(g) \longrightarrow 2F^-(aq)$	2.87
$5e^- + 8H^+(aq) + MnO_4^-(aq) \longrightarrow Mn^{2+}(aq) + 4H_2O(l)$	1.51
$2e^- + Cl_2(g) \longrightarrow 2Cl^-(aq)$	1.36
$6e^- + 14H^+(aq) + Cr_2O_7^{2-}(aq) \longrightarrow 2Cr^{3+}(aq) + 7H_2O(l)$	1.33
$2e^- + 4H^+(aq) + MnO_2(s) \longrightarrow Mn^{2+}(aq) + 2H_2O(l)$	1.23
$4e^- + 4H^+(aq) + O_2(g) \longrightarrow 2H_2O(l)$	1.23
$2e^- + Br_2(l) \longrightarrow 2Br^-(aq)$	1.09
$3e^- + 4H^+(aq) + NO_3^-(aq) \longrightarrow NO(g) + 2H_2O(l)$	0.96
$e^- + Ag^+(aq) \longrightarrow Ag(s)$	0.80
$e^- + Fe^{3+}(aq) \longrightarrow Fe^{2+}(aq)$	0.77
$3e^- + 2H_2O(l) + MnO_4^-(aq) \longrightarrow MnO_2(s) + 4OH^-(aq)$	0.59
$2e^- + I_2(s) \longrightarrow 2I^-(aq)$	0.53
$4e^- + 2H_2O + O_2(g) \longrightarrow 4OH^-(aq)$	0.40
$2e^- + Cu^{2+}(aq) \longrightarrow Cu(s)$	0.34
$e^- + Cu^{2+}(aq) \longrightarrow Cu^+(aq)$	0.15
$2e^- + Sn^{4+}(aq) \longrightarrow Sn^{2+}(aq)$	0.13
$2e^- + 2H^+(aq) \longrightarrow H_2(g)$	0.00
$3e^- + 4H_2O(l) + CrO_4^{2-}(aq) \longrightarrow Cr(OH)_3(s) + 5OH^-(aq)$	-0.12
$2e^- + Pb^{2+}(aq) \longrightarrow Pb(s)$	-0.13
$2e^- + Sn^{2+}(aq) \longrightarrow Sn(s)$	-0.14
$2e^- + Ni^{2+}(aq) \longrightarrow Ni(s)$	-0.23
$2e^- + Fe^{2+}(aq) \longrightarrow Fe(s)$	-0.44
$2e^- + Zn^{2+}(aq) \longrightarrow Zn(s)$	-0.76
$2e^- + 2H_2O(l) \longrightarrow H_2(g) + 2OH^-(aq)$	-0.83
$3e^- + Cr(OH)_3(s) \longrightarrow Cr(s) + 3OH^-(aq)$	-1.30
$3e^- + Al^{3+}(aq) \longrightarrow Al(s)$	-1.66
$2e^- + Mg^{2+}(aq) \longrightarrow Mg(s)$	-2.37
$e^- + Na^+(aq) \longrightarrow Na(s)$	-2.71
$2e^- + Ca^{2+}(aq) \longrightarrow Ca(s)$	-2.87
$e^- + K^+(aq) \longrightarrow K(s)$	-2.93

APPENDIX D

SPECIALIZED TECHNIQUES

D.1 VOLUME MEASUREMENTS

There are several pieces of glassware designed to measure liquid volumes. Some give only approximate measurements (droppers, graduated beakers and flasks) while others give very precise measurements (graduated cylinders, pipets, burets). When using these more precise measurers, the experimenter must be aware of two factors; how to read a miniscus and whether the glassware is designed to hold or to deliver the measured volume.

The liquid surface in any tube will be slightly curved, not perfectly horizontal. This curved liquid surface is called a meniscus. It will be curved downwards (as with water) when the attraction of the liquid molecules to the atoms of the glass is stronger than the attraction the liquid molecules feel toward each other. With liquids such as mercury, the meniscus is curved upwards because the Hg atoms feel a stronger attraction for each other.

When measuring the level of a liquid, we must decide which part of the meniscus to read. Since consistency is the important consideration, chemists read the rounded portion of the meniscus (i.e., the bottom of the meniscus for H_2O solutions). When reading a liquid level, the eye must be **on the same horizontal level** as the meniscus. This is to avoid the problem of parallax, choosing the wrong measuring line due to an improper viewing angle.

Reading a Meniscus Level

16 - 20 drops = 1 mL

267

The more precise measuring glassware is marked to measure either the amount of liquid being **held** by or the amount of liquid that can be **delivered** (extracted) from the glassware. No one volume marking can precisely measure both of these because all the liquid cannot be poured from a container. Some will always remain behind, clinging to the container walls. The experimenter should be aware of which glassware is marked to measure liquid held or liquid delivered and use this glassware accordingly.

A. DROPPERS AND GRADUATED BEAKERS

These are used to give approximate measures of volumes. The dropper is the quickest way to measure volumes of less than a few milliliters. The measurement is made by counting free falling drops taking 1.0 mL = 18 drops. The method is precise to ±10% as one mL may be equal to 16 to 20 drops depending on the particular dropper used. For larger volumes the markings on beakers and flasks can be used. These are accurate to only ±5% because of the large diameters of the glassware and the methods of manufacture. It is not important here whether the volume marking refers to the volume held or delivered since the inherent error is so large.

B. GRADUATED CYLINDERS

The volume markings here measure the amount of liquid held by the cylinders. Their accuracy is about ±1 to 2%. The error is much larger, of course, if the cylinder is used to measure liquid delivered. One way around this problem is to do some calibrations ahead of time. For example, assume you wish to use a graduated cylinder to deliver about 100 mL of solution several times in an experiment. Take two 100 mL cylinders and fill one with water to the 100.0 mL mark. Now pour this water gently into the second, dry, cylinder. Read the volume delivered. Within the error limits of ±1 to 2%, this is the volume the first cylinder will deliver throughout the experiment. (This assumes that your pouring procedure will be the same during the experiment.)

C. PIPETS AND BULBS

Pipets are used to deliver measured volumes with an uncertainty of only about ±.02% and should be used whenever highly accurate work is desired. There are two general types; volumetric pipets which have only one volume mark and graduated pipets which have several marks. The pipetting bulb is used to draw the liquid into the pipet. (There is the temptation to disregard the bulb and draw the liquid into the pipet by mouth suction. Do not give in to temptation! Always use the bulb! Mouth suction can easily lead to ingestion of corrosive or poisonous chemicals.) The proper use of the pipet and bulb requires a fair amount of skill. It may be worth the time to practice and to perfect your skills using distilled water.

Preparing the Pipet. In order to not contaminate or dilute the solution being measured, the pipet must be cleaned and its insides coated with the solution to be measured.

1. Draw distilled water into the pipet until about 1/5 of its "fat" portion is filled. Take off the bulb, quickly placing your thumb or forefinger over the pipet end. This will keep the water from draining out. Keeping your finger in place, tip the pipet to a horizontal position. Rotate the pipet so that the entire "fat' portion is repeatedly coated with water.

   Now allow the water to move down the "bulb" stem until it is about 2/3 of the way between the volume mark and the end of the stem. This is done by tipping the pipet to about a 45 degree angle and slightly releasing your finger pressure. (It is wise to not let any liquid reach the end of the "bulb" stem. If the end is wet it will be almost impossible to control the flow of liquid with finger pressure during the actual measurement.)

   Tip the pipet back to the normal vertical position ("bulb" stem up) and let the water drain into a sink. Check the coating of the inside. If it is uniform, having no breaks or localized droplets formed, proceed to Step #3.

2. An uneven coating with water indicates that some insoluble chemicals, such as "oils", are still adhering to the glass. Wash the pipet 2 to 3 times with a mild soap solution using the procedure outlined in Step #1 above. Then rinse out the soap solution with 4 to 6 treatments of tap water.

graduated pipet

volumetric pipet

volume mark

Rinsing

3. Using the procedure outlined in Step #1 above, rinse the pipet twice more with distilled water. Then rinse it twice with the solution to be measured. This will replace the water coating with a coating of the solution.

Measuring Procedure. The inside of the pipet should be coated with the solution to be measured as described above. If you will be using the same pipet to measure several solutions of the same chemical at different concentrations, it is advisable to start with the one of lowest concentration.

4. Attach the bulb to the pipet stem to a depth of no more than a few millimeters. (If the stem is inserted too deeply into the bulb, it will be difficult to separate the two later.) Compress the bulb and insert the tip of the pipet into the solution. Slowly release the pressure on the bulb allowing the solution to be drawn up the pipet. Continue releasing the bulb pressure until the liquid level is about half way between the volume mark and the bulb.

5. Keeping the pressure on the bulb constant, push the bulb off the stem with the thumb or forefinger of the other hand. In the same action, quickly clamp the thumb or forefinger over the stem end. This will hold the liquid level above the volume mark.

push
with
finger

cap hard
with finger

Removing the Bulb

Draining a Pipet

6. Wipe the pipet tip with a clean towel. Release your finger pressure ever so slightly allowing the solution to drain **just to** the volume mark. This can be done by slightly rotating the finger on the stem or by slowly relaxing the tension in the finger. The draining solution can be caught in the beaker holding the remainder of the solution. Touch the tip of the pipet to the beaker to remove any partially formed drops.

If the liquid level falls below the volume mark or if any air bubbles are present in the pipet, drain the solution back into the original beaker and start over.

7. Hold the pipet tip against the inside glass of the flask that is to receive the measured volume. Remove the finger from the stem and allow the solution to drain from the pipet. After drainage, touch the pipet tip to the inside glass of the flask or to the surface of the solution. This will remove the last required amount. There will be a small amount of solution still in the pipet tip, **do not** blow this out. The pipet has been calibrated to leave this amount undelivered.

8. If you are to use the pipet to measure a solution of the same chemical but at a different concentration, rinse the pipet twice with the new solution before measuring. If a solution of a different chemical is to be measured, rinse the pipet 3 times with distilled water than 2 times with the new solution. When you are through with your pipet work, rinse it 3 to 5 times with distilled water. Use the rinsing directions outlined in Step #1 above.

D. VOLUMETRIC FLASKS

Volumetric flasks are used to hold measured volumes. Usually they are used to prepare solutions of known molar concentrations. The flasks are precise to $\pm$ .02%. One must be concerned with contamination as well as precision in using the flasks.

In the following instructions, it is assumed that distilled water is to be the solution solvent. If this is not the case, use the desired solvent where distilled water is called for.

1. Rinse the flask 3 times with distilled water. This can be done by putting a small amount (about 1/5 the flask volume) of water in the flask and then rotating the flask until the entire "fat" portion and stem have been repeatedly coated.

   If you note that the water coating of the flask is uneven, the flask is dirty and must be cleaned. Put a small amount of soap solution in the flask. Shake and rotate the flask until the entire inside is coated. Repeat this treatment twice more with fresh soap solution. Rinse the flask 4 to 6 times with tap water to remove all traces of the soap. Then rinse 3 times with distilled water.

2. Quantitatively (precisely) transfer the solute (the material to be dissolved) into the flask. If the solute is liquid, measure it into the flask using a pipet or a buret. If it is a solid, accurately weigh the solute into a small beaker or onto a creased piece of weighing paper. Hold the beaker or paper against the rim of the flask opening. Tip the beaker or paper to a slight angle and use a spatula to scoop small smounts of solid at a time into the flask. Finally rinse the spatula and the beaker or paper with distilled water catching **all** the rinse in the flask. These things should be rinsed at least 3 times to insure that all of the solute is in the flask.

3. Add distilled water until the "fat" portion of the flask is about 4/5 full. Swirl the flask to thoroughly dissolve the solute.

4. Carefully add distilled water until the liquid level is **just at** the volume mark. Stopper the flask. Mix the solution by inverting the flask. This is done by quickly turning the flask upside down allowing the air gap to move through the solution. Mix by inversion about 7 times. Be sure to hold the stopper firmly in place. (Mixing by shaking simply does not work well with volmetric flasks.)

   After mixing and removing the stopper, you will often notice that the liquid level is slightly below the volume mark. This is simply because some of the solution is now coating the upper stem and stopper. Do not be concerned and **do not** add more distilled water.

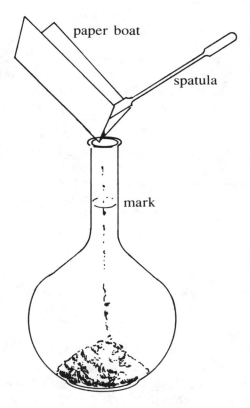

paper boat

spatula

mark

**Adding a Solid to
a Volumetric Flask**

air bubble

hold in
stopper

Mix by Inversion

E. BURETS

Burets are calibrated to measure the volume of liquid delivered and are precise to $\pm .02\%$. Burets are used whenever odd volumes are to be delivered or more commonly, in titration analysis experiments.

The purpose of a titration is to determine the precise volume of a solution needed for a reaction. This is done by using the buret to add small amounts of the solution to the reaction flask until some indicator changes color. The amount of solution added is determined by noting the buret volume readings before and after these additions. The solution in the buret is called the **titrant**. The point in the experiment where the indicator permanently changes to the desired color is called the **end-point**.

Preparing the Buret. In order not to contaminate or dilute the solution being measured (the titrant), the buret must be cleaned and its insides coated with the titrant before use.

1. Rinse the buret 3 times with distilled water. This can be done by filling the buret 1/5 full with distilled water and then tipping and rotating it to coat the entire insides with the water. Next hold the buret upright and turn the stopcock allowing the water to drain into a "refuse" beaker. Note the coating of water on the buret walls.

2. If the water coating is uneven, showing breaks or droplets, the buret is dirty and should be cleaned with soap solution. Add some soap solution and clean the entire barrel using a buret brush. If brushes are not available, rinse the buret 3 times with soap solution using the procedure outlined in Step #1 above. Be sure to drain some soap solution through the stopcock and tip.

   Rinse the buret barrel, stopcock and tip about 5 times with tap water to remove all traces of soap. Then rinse these parts 3 times with distilled water.

3. Using the procedure in Step #1, rinse the buret barrel, stopcock and tip at least twice with the titrant solution. This will replace the water coating with a coating of titrant.

Titration Procedure

4. Secure the buret to a stand. Fill it with titrant to a point above the zero volume mark. Place the refuse beaker under the buret and turn on the stopcock full force allowing the titrant to drain a bit below the zero mark. This full force drainage is meant to eliminate any air gaps in the stopcock and tip. Check the tip and barrel for air gaps and bubbles. If there are any, consult with your instructor.

5. Record the volume reading on the buret. (Your initial volume reading need not be at zero.) Remove any drops on the tip by touching the tip to the side of the refuse beaker.

6. Place the solution flask to be titrated under the buret. Be sure the indicator has been added to the flask. By carefully turning the stopcock, add **small amounts** of titrant to the flask. Swirl the flask after each addition to completely mix the two solutions.

begining
level
below
zero

barrel

stopcock

tip

Swirl After Each Small Addition

7. Continue additions until one drop of titrant causes a permanent (30 sec) color change in the swirled flask. This is the end-point of the titration. Record the volume reading of the buret and calculate the total volume of titrant solution that has been added.

Pointers on Titrating. The following are some "tricks of the trade" that can improve the accuracy of your work.

8. Make your buret readings as precisely as possible. Determine the readings to a hundredth of a milliliter by estimating the distance between volume markings. Wait 15 sec before making a reading. This allows the solution on the sides of the barrel to finish draining. If you find it difficult to see the exact bottom of the miniscus, place a darkened piece of paper (see figure) behind the buret and about 1 cm below the meniscus. The meniscus will "reflect" this darkening and become more distinct.

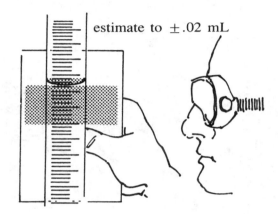

estimate to ±.02 mL

Making the Meniscus Clearer

9. One should locate the end point within 1 drop (1/20 mL). Adding one drop at a time through the entire titration takes a great deal of time and is boring. There is a way around the problem. Begin by adding about 1/2 to 1 mL at a time and watch the coloring where the titrant hits the solution. If the color change dispels by itself, the end point is not close. If you must swirl the flask to dispel the color change, you are getting closer and must cut down the size of the addition. When 2 or more swirls are needed to dispel the color change, you are very close and should make single drop additions. Alternatively, if you have a reasonable idea of where the end point should be, add 80% of the needed titrant in one shot and then begin your dropping additions.

10. Place a white sheet of paper under the flask. This will allow you to see the color changes more clearly.

11. Near the end of the titration, rinse down the insides of the flask with a little distilled water to return all chemicals to the body of the solution.

12. Do not trust the results of a single titration. If possible, repeat the titration of a particular solution a couple of times to check on the reproducibility of your result.

D.2 SPECTROPHOTOMETER MEASUREMENTS

A spectrophotometer is an instrument that measures the color intensity of a solution. The color intensity can then be related to the concentration of the solution. The basic principles involved are used by each of us in our everyday lives. For example, we judge the strength of tea or coffee by its color. The darker the color, the stronger the drink. We can rephrase this to say; the more intense the color, the higher is the solution concentration. The spectrophotometer does just what your own eyes do: acts as a color intensity meter. The advantage of the instrument is that its measurements are much more precise than those of your eye.

Most spectrophotometers have two scales for measuring color intensity: an absorbance scale (A) and a percent transmittance (%T) scale. The two scales are related mathematically as shown later. However the absorbance (A) readings are the most useful because they are directly proportional to the molar concentration of the solution.

$$A = (constant) \times (concentration)$$

The constant in the equation depends primarily on the chemicals used in the solution (solute and solvent) and on the wavelength setting on the instrument. Normally the value of the constant is calculated from absorbance readings of solutions whose concentrations are known precisely. This value is then used with similar solutions whose concentrations are unknown.

A more detailed discussion of how a spectrophotometer works is given in part B. It is suggested that you study this section both before and after you use the instrument in lab. This will help put your data in perspective.

A Typical Student Model Spectrophotometer

A Typical Cell (sample holder)

A. PROCEDURE

These are general procedures for making precise measurements. Study them and incorporate any changes given by your instructor for your specific spectrophotometer.

There are three things the experimenter must watch if precise measurements are to be obtained (1) that the instrument is standardized; i.e., gives proper readings at each end of the scale, (2) that the solutions measured are not contaminated or diluted, (3) that the outside of the cell is spotlessly clean and dry.

1. Turn on the instrument and adjust the wavelength knob to the required setting. Allow the instrument to warm-up for the required amount of time.

2. Rinse the cell 4 to 6 times with distilled water. Fill it about 3/4 full with water. Holding the cell at its top, wipe the entire outside dry with tissue or a clean towel. Check the outside for smudges and liquid. If any exist, remove them. From now on handle the cell only at its top.

3. With no cell in the instrument set the reading at exactly 0%T (A = ∞) using the proper knob.

4. Insert the water cell into the cell compartment so that the mark on the cell is properly aligned. Close the compartment lid. Using the proper knob adjust the reading to exactly 100%T (A = 0).

5. Repeat steps nos. 3 and 4 until proper readings are obtained at both 0%T and 100%T.

6. Rinse the inside of the cell at least twice with small portions of the solution to be measured. (The purpose of this is to prevent dilution by coating the entire inside of the cell with the solution.) Fill the cell about 3/4 full with the solution. Holding the cell at its top, rinse the lower 3/4 of its outside with distilled water. Wipe the entire outside of the cell removing all liquid films and smudges.

7. Holding the cell at its top, insert it into the cell compartment. Align it to the mark and close the compartment lid. Once the reading stabilizes, record it as directed (A or %T).

Other solutions can now be measured by repeating steps nos. 6 and 7. It is wise to check the standardization of the instrument (Steps nos. 2 to 5) about every 20 minutes. If you have enough solution, it is also recommended that you make 2 or 3 separate measurements of each solution to obtain an average reading.

Absorbance Versus Wavelength

B. DETAILS OF LIGHT ABSORPTION

The reason why some solutions are colored is that selected portions of the light passing through them are absorbed (captured) by the solute molecules. Several factors affect the portion of light and the fraction of light absorbed.

Wavelength. Light, also called electromagnetic radiation, is pure energy. It may be viewed as packets of energy, called photons, traveling with a wave motion. The energy (E) of each photon gives its motion a characteristic wavelength (λ). The relationship is $E = hc/\lambda$ where h is Plank's constant (6.62×10^{-34} joule - sec.) and c is the speed of light (3.00×10^8 m/sec). Thus light of longer wavelength has photons of lower energy.

A molecule will absorb light whose energy corresponds to one of its own energy needs. Thus some wavelengths of light are heavily absorbed while others remain untouched. The variation in absorbance with λ is illustrated for two ions on page 276. The visible spectrum is light having λ between 400 and 800 nanometers ($1nm = 10^{-9}m$). Our eyes "see" these different wavelengths as different colors. (Light having all these wavelengths is clear or "white" light.) When looking at a solution we see only those colors (wavelengths) not absorbed by the solute molecules. Thus a Cu^{+2} solution is blue because yellow, orange and red wavelengths have been absorbed.

Nature of the Chemicals. Visible light has energy matching that needed to move electrons to higher energy levels in many molecules. However since the electron energy levels in each type of molecule have slightly different spacings, each chemical species will show a different pattern of absorbance versus λ .

Concentration of the Solution. As the concentration rises the light photon has a greater chance of encountering an absorbing molecule as it passes through the solution. Thus the fraction of light absorbed increases with higher concentrations. The absorbance also increases with the **Length of the Cell** for much the same reason. If the light photon must pass through more solution, the probability that it will be absorbed increases.

In a light measuring experiment, the experimenter must have a method of controlling these factors if the data are to be useful. The spectrophotometer is designed to give such control. A simple schematic of a spectrophotometer is given below. White light from a high intensity lamp is focused on a grating or prism which fans out the light according to its various wavelengths. The grating can be rotated so that light of just the required wavelength hits the sample. The light passes through the sample and hits a light detector, called a photomultiplier. The detector converts the light intensity into an electrical current which moves the needle over the reading scale.

Schematic of the Instrument

277

In standardizing the instrument, the needle is set to 0%T (A = ∞) with no light reaching the detector, i.e., intensity I = 0. It is set to 100%T (A = 0) when the detector senses I_{max}, the light intensity emerging from the cell and water. This is the maximum amount of light that can emerge from any sample. If the solution sample is now inserted, the scale will read a light intensity, I, that is properly compared to the no light and to the maximum intensity situations.

I = 0

partition

I

sample

I_{max}

water

**The Two Standardization and
the Sample Readings**

The mathematical relationship for light absorption is as follows:

$$\%T = \frac{I}{I_{max}} \times 100$$

Thus the percent transmittance measures the ratio of the intensity emerging from the solution to the intensity emerging from water alone. It is an indication of the fraction of light captured by the solute. The higher the fraction of light captured the lower the %T.

Unfortunately these intensity fractions do not relate simply to concentrations. They are related through the common log function (i.e., see pH's).

$$-\log\frac{I}{I_{max}} = E \times d \times c$$

c is molar concentration of the solution, d is the length of the light path through the solution (i.e., the inner diameter of the cell) and E is a constant for the particular solute and wavelength used. E is called the extinction coefficient.

Because finding log's of numbers is tedious, the reading scale of most spectrophotometers are marked to give the log function directly. This scale reading is called Absorbance, A.

$$A = -\log \frac{I}{I_{max}}$$

or $$A = E \times d \times c$$

or $$A = (constant) \times c$$

It must be remembered that this last equation holds only when E and d are constant. That is, where the absorbing compound, the wavelength and the diameter of the cell are not changed.

The numerical value of the "constant" can be determined as follows. Make solutions of precisely known concentrations of the compound in question. Several should be made, each with a different concentration. Measure the absorbance of each solution by the procedure outlined above. Plot the data as A versus c and draw the best straight line through them. The slope of the line is the "constant." Alternatively you can determine A/c for each known solution and average the results to get the best value.

D.3 pH AND CONDUCTIVITY

A. pH

The Concentration of H$^+$ in moles/liter in water solution, [H$^+$], is controlled by the following reaction:

$$H_2O \rightleftharpoons H^+(aq) + OH^-(aq).$$

This reaction does not proceed to completion, and the concentrations of [H$^+$] and [OH$^-$] are very small. Because [H$^+$] generally have values between 10^{-1} and 10^{-14}, it is convenient and more compact to express these numbers as logarithmic functions of the H$^+$ concentration. Since these numbers are less than one, their logarithmic function will be negative, thus it is also convenient to change the sign. This compact method of expressing H$^+$ concentration is called pH and is defined by the following equation:

$$pH = -\log[H^+].$$

In order to determine the pH of a solution determine the concentration of H$^+$ in moles/liter, take the logarithm and multiply the number (usually negative) by -1.

In order to determine the [H$^+$] of a solution given the pH, multiply the pH by -1 and take the antilog of the number : [H$^+$] = 10^{-pH}.

The pH of a solution can be approximately determined using pH paper or indicators, or more exactly by using a pH meter.

B. THE pH METER

Schematic of the Instrument

A pH meter consists of two electrodes which generate an electrical signal that can be read on a meter in pH units. One of the electrodes is sensitive directly to [H+]. It is called a glass electrode and it has a very thin spherically shaped glass membrane at its tip. The other is a reference electrode which completes the electrical circuit. Reference electrodes will clog up if allowed to dry out. If this occurs see your lab instructor about what to do. A complete circuit depends on both electrodes having their tips completely submerged in the solution to be measured.

Your pH meter may be equipped with a combination electrode which has both electrode systems in a single probe. These two electrodes are still connected to each other externally through the glass electrode tip and a small pin hole or wick on the side of the tube. Make sure the probe is covered with solution to depth sufficient to cover the pin hole.

MAINTENANCE AND CARE OF pH METERS AND ELECTRODES

The electrode tip is the most delicate part of this instrument. Be extremely careful that **nothing** touches it except liquids and tissue for drying. Do not allow it to touch the bottom of a beaker. When drying it with tissue, **DO NOT RUB DRY**, PAT IT DRY GENTLY.

Each pH meter should come equipped with the following auxiliary items: one box of tissue, one squeeze bottle of distilled water, one container of buffer solution, one container of soaking solution and one large beaker to collect waste wash solution.

At the beginning of each lab, check the meter for these items:

1. The power plug should be firmly seated and grounded in the socket. If the meter needle sways wildly (in the pH mode) when you pass your hand over the meter, the ground is not secure.

2. The electrode input jacks should be firmly seated all the way in their sockets.

3. Check the electrodes for cracks. Cracked or severely scratched electrodes do not work.

4. The electrodes use liquid junction references. Therefore, the reference electrodes and the outer jackets of combination electrodes must be kept full with saturated KCl. If the level is low, fill through the hole in the body below the cap until about 1/4 inch below the hole. The holes should remain open. Some electrodes do not require filling.

5. Buffer solutions should be checked with pH paper to see if they are still reliable. If there is any doubt, get fresh buffer.

6. Electrodes should be stored in a soaking solution when not in use.

7. Report non-functioning meters to your lab instructor.

INSTRUCTIONS FOR MEASURING pH

Most pH meters have similar operating procedures. The only differences will be the position of switches and dials.

1. Check that the temperature dial is set at 20°C and the function dial is on Standby.

2. Move the support up so that the electrode tip is out of the soaking solution. Set this solution aside.

3. Using the squeeze bottle, rinse the bottom third of the electrode, catching the water in the waste beaker. Gently **PAT** away the excess water from the electrode using a clean tissue.

4. Lower the electrode into a beaker of buffer solution. Only one inch of the electrode need be immersed. Be careful to not bump the electrode tip on the sides or bottom of the beaker.

5. Move the function switch to pH. Turn the Standardize dial until the meter reads the pH of the buffer solution.

6. Move the function switch to Standby.

7. Repeat steps 2 and 3 to again rinse the electrode. Set aside the buffer solution and be sure not to contaminate it.

8. Lower the electrode into the sample to be measured. Turn the function switch to pH and record the reading. Return the function switch to Standby.

9. Repeat steps 2, 3, and 8 to obtain reading of other samples.

10. When readings are complete, rinse the electrode as in steps 2 and 3. Immerse the electrode in the soaking solution (0.01 M HCl). Be sure the function switch is on Standby.

C. THE CONDUCTANCE METER

Conductance measures the ease with which electrical current can pass through a solution. By definition, conductance is the reciprocal of the electrical resistance of the solution (common units are mhos or μS, micro-Siemens).

Pure water does not conduct electricity, only solutions with ions dissolved in them will conduct. The higher the charge on the ions dissolved and the higher their concentration, the higher will be the conductance of the solution. Thus conductance measurements can be used, qualitatively, to tell the difference between types of ions (for example, +1 versus +2 cations) or to tell the difference between two concentrations of the same salt.

Several kinds of conductance meters are available; two of the easier ones to use are shown below. Your instructor will give specific directions for using your meter, but they will generally follow this order:

1. Shut the meter off. **DISCONNECT THE POWER CORD ON THE LIGHT BULB APPARATUS**.

2. Rinse the electrodes with distilled water, and pat them dry.

3. Immerse the electrodes in the solution to be measured.

4. Turn the meter on. Wait a few seconds for the reading to stabilize, and record the reading.

5. Turn the meter off **(DISCONNECT THAT POWER CORD)** before pulling the electrodes from the solution. Repeat steps 2 to 4 for other solutions.

Light Bulb Apparatus

Digital Conductance Meter

D.4 QUALITATIVE ION ANALYSIS USING SEMIMICRO TECHNIQUES

In these experiments you will be concerned with determining what particular ions are in a solution or solid. This concern with only what is present is called qualitative analysis. (A concern for how much is present is called quantitative analysis). You will be working with very small amounts of solutions, usually about 1/4 to 1/2 of a milliliter. Thus the experimental procedures are called semimicro (small amount) techniques. Two other definitions of importance are:

> **precipitate**—a solid formed when two solutions are mixed. (abbreviated ppt.)

> **supernatant**—the liquid phase of a solid plus liquid mixture. (abbreviated spn.)

The basic idea in these experiments is to add a reagent to a solution of ions and to note any and all changes that occur. In particular one looks for and notes down

1. the formation and color of a ppt.

2. changes in color of the solution or spn.

3. redissolving of a ppt.

4. evolution of a gas.

A. ADDITIONS AND MIXING.

The amounts of solutions to be mixed in each test are often critical to the results and must be carefully watched. The amounts are usually controlled by counting drops of solution used (18 drops = 1 mL). When making additions suspend the end of the dropper just above the opening of the test tube and gently apply pressure to the dropper bulb so that the drops fall one at a time into the test tube.

When test reagent has been added to an ion solution, mix the two completely before you note down your observations. The results that you see when one solution is layered on top of the other may not be the same as when they are mixed. To mix the solutions move a clean stirring rod gently up and down within the test tube; i.e. use a "butter churn" type motion. Watch that the liquid does not overflow the test tube.

18 drops ≅ 1 mL   "Butter Churn" Mixing

B. THE CENTRIFUGE

The centrifuge is a machine that spins a test tube solution at high speeds causing the heavier and lighter parts of the mixture to separate. In essence, it accelerates what gravity would have done anyway. The centrifuge is used to separate a precipitate from the supernatant. Usually only one or two minutes of centrifuging are needed to separate the ppt. and spn.

Precautions to be observed in using the centrifuge are as follows. (1) Do not have the test tubes more than 4/5 full. This will prevent spilling in the centrifuge. (2) Place the tubes in the centrifuge in a symmetrical pattern such as one tube directly across from another. If you are centrifuging just one test tube, place another containing water across from it. If the placement of the tubes is not balanced, the spinning will be uneven and the centrifuge will move ("walk") on the bench. (3) Allow the centrifuge to slow down by itself. Helping it slow down can cause injury. Also if the slow down is too abrupt the ppt. and spn. will remix.

Effect of Centrifuging   **Decant by Pouring or Drawing**

C. DECANTING

This is the term used to denote removing a supernatant from a precipitate. There are two decanting methods. The first is to simply pour the spn. out of the test tube leaving the ppt. behind. The second method is to draw the spn. out of the test tube using a dropper. This dropper method is used when the ppt. is not a firm pack and might pour out with the spn. It is also the preferred decantation method where the spn. is to be saved for later use. It is often difficult to pour directly from one small test tube to another.

When using the dropper method be sure to compress the bulb **before** inserting the dropper. A rush of bubbles in the spn. will cause the ppt. and spn. to remix.

D. WASHING A PRECIPITATE

Not all of the supernatant is removed from the test tube when one decants. A few drops remain on top of the precipitate, some clings to the test tube walls and some is trapped within the packed precipitate. If a later test on the ppt. requires that the chemicals in the spn. be absent, these traces of spn. are removed by a procedure called washing.

Washing a ppt. involves the following sequence of steps. (1) Decant the original spn. (2) Add to the ppt. about 1 mL of distilled water (or other prescribed liquid). (3) Completely mix the H_2O and ppt. This allows the H_2O to leach out any spn. trapped in the packed ppt. (4) Centrifuge the mixture. (5) Decant the liquid, discarding it. (6) Repeat the process. Usually two such water washes are sufficient as they will leach out about 97% of the spn. originally left in the test tube.

Washing a ppt.

E. HEATING

You should **not** use a burner flame to directly heat liquid in a small test tube. Such direct heating is too intense and too localized. It will usually cause the liquid to blow out of the test tube. This can cause injury and, in any case, ruins the experiment.

Heating is to be accomplished by resting the test tubes in beakers of heated water. This heating is more even and keeps the temperature from exceeding 100°C. If a test tube heating platform (also called an aluminum holder) is available it should be used in the beaker. Otherwise use a beaker narrow enough to keep the test tubes protruding over the beaker lip.

Adjust the water level in the beaker so that only the lower 1/2 to 3/4 of the test tube is submerged. Move the burner flame in and out as needed to keep the water hot but to prevent violent boiling.

Heating a Sample

Testing pH

F. pH MEASUREMENT AND CONTROL

It is important in many reactions that the solution's pH be carefully controlled. (See Appendix D.3 for an explanation of pH.) The control is achieved by adding an acid or base solution dropwise until the required pH is obtained. The pH of the solution should be measured after each dropwise addition by using litmus or pH paper. Do not dip the pH paper into the solution. This can lead to chemicals from the paper getting into the solution. The proper procedure is to mix the solution with a stirring rod and then to touch the wet end of the rod to the test paper.

The pH of a gas bubbling out of a solution can be tested by wetting the test paper with water and then holding it over the mouth of the test tube.

G. CONTAMINATION

Even the smallest contamination of your ion or reagent solutions can change the test results. Contamination is a particular problem with these experiments because you reuse equipment so often. **Clean and distilled water rinse each piece of equipment after each use**. Thus you should clean a dropper before using it in a different solution, clean a stirring rod before using it to mix a second solution, etc.

Most often cleaning requires only that the test tube, rod, etc. be rinsed several times with tap water then 2 to 3 times with distilled H_2O. The equipment does not have to be dry. Use a test tube brush to clean out stubborn ppts. A dropper can be cleaned by removing the bulb and running water down the barrel.

more
reagent

Testing for Complete Precipitation

H. COMPLETE PRECIPITATION

In Section Two of each "L" experiment you will be analyzing a solution containing more than one ion. (You will create an analysis scheme—a game plan—for this work.) In this analysis of an unknown you will often depend on precipitating an ion completely from solution to separate it from the other ions. It is important that this precipitation reaction be complete.

After centrifuging the ppt. plus spn. mixture, add a drop or two more of the test reagent. If new ppt. forms in the spn., add two more drops of test reagent, mix and recentrifuge. Again add a drop of reagent. Continue this 1 to 2 drop testing of the spn. until no new ppt. forms. At this point the reaction is complete.

I. ORGANIZATION AND REAGENTS

It is strongly suggested that you take a few minutes to plan your work, especially when doing the Section One experiments. You will be performing 25 or more individual tests in one period; thus, some organization is needed to save time and to not confuse your results. The following ideas might be helpful.

1. When taking chemicals from the stock bottles, immediately label the beaker used with both the reagent solution's **name** and **concentration**. Often you will have to use the same chemical at different concentrations at different times. Take only the small amounts of stock reagents needed (usually about 5 mL).

2. After using a dropper, immediately return it to its reagent beaker. Do not have two droppers out of their reagent beakers at one time. This will keep you from returning a dropper to the wrong reagent beaker.

3. Label each test tube and assign it a labeled position in your test tube rack. Never have two test tubes out of the rack at the same time, always return one before picking up another. This keeping track of test tubes is needed because the label on the tube can be smeared in handling. Be sure the label on a tube is clear before you put it into a centrifuge.

4. Put all used, dirty glassware in one section of the bench so as not to confuse it with clean glassware.

5. When you are to run a test on several different ion solutions, run the test on all of them simultaneously. Set up a labeled tube for each ion solution in the rack. Put the proper ion solution into each test tube. Next go from one to the other adding the test reagent and mixing. This simultaneous testing saves time and allows you to make direct comparisons of the results for different ion solutions.

6. Both lab partners should observe the results of each test and record their own perception of the results. Also describe fully your observation, do not simply put down "Reacted." You will need to recall exactly what you saw happen when you analyze your unknown.

SAMPLE ANALYSIS SCHEME OR FLOW CHART

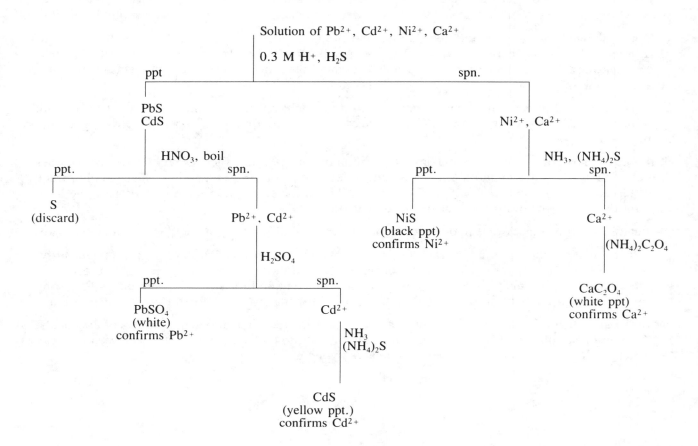

D.5 FILTRATIONS

A large number of chemical reactions in solution result in the formation of an insoluble solid. It is sometimes necessary to weigh the quantity of solid (called a precipitate) to determine the amount formed. In order to do this it is necessary to first separate the precipitate from the liquid phase of the mixture (called the supernatant liquid), and then to dry the precipitate residue. One of the simplest and most useful methods for separating such a two phase mixture is by filtration.

GRAVITY FILTRATION

1. Clean and rinse a filtration funnel with distilled water.

2. Obtain a piece of filter paper of the proper size and grade (fine precipitates require finer grade paper) and prepare it according to the following:

| **fold in half** | **fold again so that. . .** | **the top fold is slightly smaller** | **Optional—tear off corner of smaller fold (to give a better seal)** | **open cone so that three layers (with torn cover) are on one side** |

3. If the filter paper needs to be weighed, do so after the corner is torn off and before wetting it.

4. Place the cone in the damp filter funnel and wet with distilled water. Gently press the upper edge of the filter paper against the funnel to form a tight seal. The funnel stem should hold a column of water without bubbles. This condition allows maximum speed in filtration.

5. Gently pour the solution to be separated into the filter being careful not to overflow the top of the filter paper.

6. Wash the residue with distilled water. If there is a danger of losing some of a partially soluble residue wash with small amounts of the supernatant liquid or another liquid which will not dissolve the residue.

7. Sometimes a precipitate may be too fine to be caught in the filter paper. If the precipitate passes through use one of the following procedures.

    a. Filter the supernatant with passed precipitate again. Often the residue in the filter paper will act to filter the passed precipitate during a second or third filtration.

    b. Use a finer grade filter paper.

    c. Use a double thickness of filter paper. Don't forget to add the weight of the second piece of paper if your experiment requires weighing.

    d. Digest the precipitate by boiling it for a period of time (20 minutes) before filtering. This increases the size of the precipitate crystals making them easier to filter. Before carrying out this procedure, check with your instructor to make sure heating your precipitate will not destroy your experimental results.

SUCTION FILTRATION

If the gravity filtration of a precipitate is particularly slow, suction filtration using a water tap aspirator can be tried. Consult with your instructor for approval and advise. A possible equipment set up is shown below.

Two disadvantages of aspirator filtrations are that they are cumbersome to use, and the supernatant might get contaminated with tap water. Some precautions to observe are listed below.

1. The filter paper should lie flat in the Buchner funnel and must cover all the holes.

2. The aspirator should be turned on **BEFORE** it is connected to the filter flask.

3. When the filtration is complete, the tubing should be disconnected from the filter flask **BEFORE** the aspirator is shut off.

DRYING THE RESIDUE

Once the mixture is filtered and washed there are several methods that can be used to dry the residue. The residue can sometimes be rinsed with a volatile solvent like acetone which will wash the water away and evaporate faster than water. Care must be taken however that the residue is not dissolved or washed away with the solvent. The filter paper should be carefully removed from the funnel and unfolded and placed on a paper towel or other absorbent material. The residue should be spread on the paper being careful not to lose any. The residue can be dried in a number of ways.

1. Air-drying. Allow the residue and paper to stand in a well ventilated space (if possible) for a period of time (overnight).

2. Hot water bath. If the residue will not decompose at 100°C, place it and the filter paper in a tared watch glass. Place the watch glass on a large beaker half filled with water. Boil the water for 10-15 minutes. Remove the watch glass and allow to cool.

3. Drying oven. If the residue will not decompose at the temperature of the drying oven, place it and the filter paper on a tared watch glass and place in the oven for 20-30 minutes.

The simplest procedure for checking if the drying process is complete is to weigh the sample, dry it again and then reweigh the sample. If the second drying causes no weight change, the sample is completely dry.

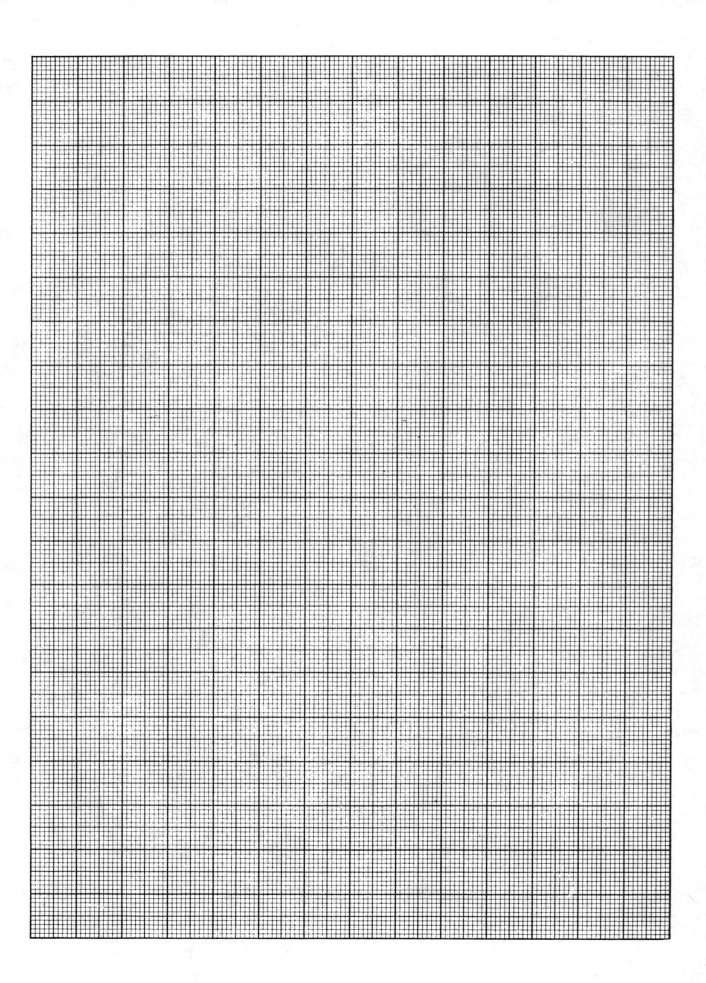